A Story of More Than 5000 Worlds

An Insight into The Possibility of Life Beyond Earth

Alejandro Ruiz Rivera, PhD

Digital: ISBN 978-1-7635654-0-1
Paperback: ISBN 978-1-7635654-2-5
Hardcover: ISBN 978-1-7635654-3-2

NATIONAL LIBRARY OF AUSTRALIA   A catalogue record for this book is available from the National Library of Australia

If you are reading a print version of this book, please be advised that an electronic version exists that links many technical terms to relevant background information on the Internet. If you are reading an electronic version of this book, please be advised that a print version of this book exists that may look better on a bookshelf.

Please note that all internet links provided in this publication were functional at the time of publishing. However, the publisher cannot guarantee the ongoing availability of these links, as their contents and addresses may change over time.

Acknowledgment of Country

The author wishes to acknowledge the Aboriginal and Torres Strait Islander peoples of this nation. We acknowledge the traditional custodians of the country throughout Australia and their continuing connection to the land, culture and community. We acknowledge the Jagera people and the Turrbal people as the Traditional Custodians of Meanjin (Brisbane), the lands on which we live and work, and where this book was written. We acknowledge the cultural diversity of all Aboriginal and Torres Strait Islander peoples and pay respect to Elders past, present and future. We celebrate the continuous living cultures of First Nations Australians and acknowledge the important contributions Aboriginal and Torres Strait Islander people have and continue to make in Australian society.

To my lovely nephew and niece Jacobo and Emma—two of the brightest stars in my night sky.

Contents

Preface

What is it about space that captivates everyone's imagination? From young kids to those who are not so young, everyone seems to be amazed by it. Is it its vastness? Is it the possibility of others out there wondering the same things we wonder about here on Earth?

My fascination with space began when my Aunt Ligia gifted me a book about the USA space shuttle at age five. She worked at a bookstore and occasionally took me there and let me pick out a new science book. I am not sure if it was the milkshake and the sandwich she would buy me after selecting a book, or the conversations with her while we were eating, what inspired my love for books and astronomy (and milkshakes). I'd like to believe that it wasn't just gluttony that drew me to science books but the curiosity that my aunt ignited in me; for that, I am forever thankful to her.

However, the book that opened my eyes to the wonders of the universe and the world of astronomy was Carl Sagan's Cosmos.[1] This book, also gifted by my aunt, is a bestseller that has inspired and continues to inspire millions of people.

During the same period, in the late 1980s, the television series 'Cosmos,' based on Carl Sagan's book, was broadcast on free-air television. I remember having lunch after school and listening to Carl Sagan's explanations of Kepler's laws, the Voyager probes, and the possibility of life— sinkers, floaters, and hunters, in the atmosphere of Jupiter.[2] Aboard his 'Ship of the Imagination', Sagan would take his audience on a journey through time and space, traversing the vastness of the universe. Often considered the greatest science communicator of all time, Sagan not only explained complex concepts in a simple manner but also had the extraordinary ability to ignite curiosity in anyone who listened.

I became an avid reader of astronomy and science communication books, and scientists and astronomers became my heroes—a sentiment that I can't hide even now when I meet a new astronomer. Given that these books often talked about the academic lives of scientists and researchers, I was one of few teenagers who knew what a Doctorate was and what it was required to earn one. Obtaining a Doctor of Philosophy (PhD) degree became one of my long-term dreams. And writing my book about astronomy was also on my bucket list (I know, I was a weird kid). I was just not sure about the specific order in which to accomplish these dreams or if they were just that.

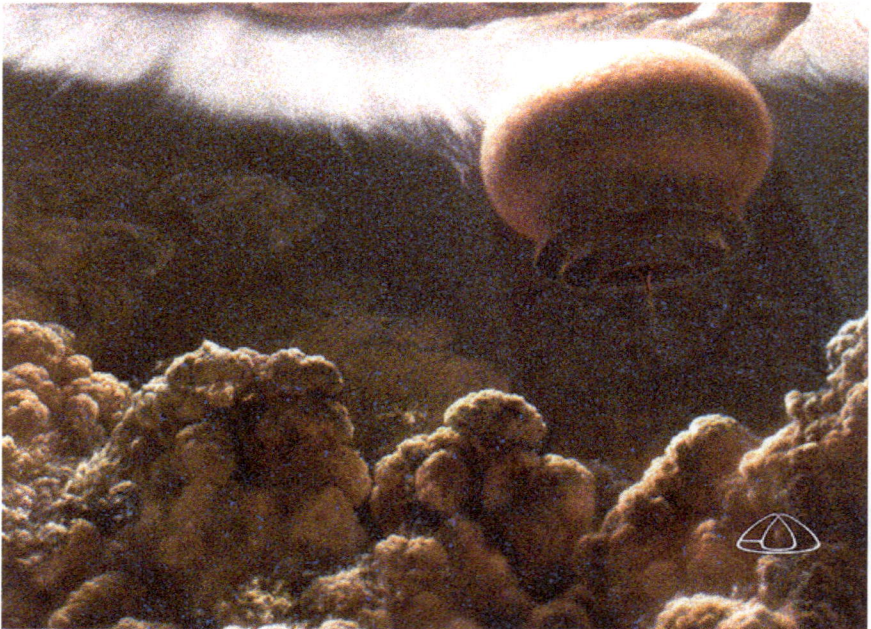

Fig. 0.1 A floater, a potential form of life in Jupiter suggested by Carl Sagan, is observed above clouds of ammonia.

But it was not only books. Since then, all the conferences (to which, when I was a kid, my dad patiently drove me all over the city of Cali), articles, documentaries, podcasts, YouTube videos, movies, or conversations I have encountered related to astronomy, and science in general, have only increased my thirst for knowledge and humbled me about our vast unknowns.

Preface

Throughout my childhood, becoming a professional astronomer was my dream. Unfortunately, in Colombia, during the 80s and 90s, as was the case with many other developing countries, one normally didn't grow up thinking that you could earn a living by pondering the wonders of the universe. Instead, I grew up considering which career would sufficiently provide for myself and my loved ones. Fortunately for Colombia, this has changed during recent years, where at least undergrads, master's degrees, and even doctorates in astronomy are available to those who wish to pursue a career in this area.

However, these options were not available to me, and I needed to find an alternative. My love for technology influenced my decision to pursue a degree in electronics engineering. Beyond astronomy, I'm deeply passionate about technology, especially its societal impact.

I treasure my university years as an electronics engineer undergrad. This particular branch of engineering requires its students to take many math and physics courses. I felt like a kid in Disneyland! I had the chance to learn about relativity, quantum physics, calculus, and linear algebra. I also treasure the many intellectual (and the not-so-intellectual) conversations I had with my classmates—many of whom I am still fortunate enough to call my friends—while trying to solve a problem or tackle a particular lab practice. Those were truly good years.

After graduation, I had the opportunity to start my professional career in the tech sector. In 2007, my wife and I decided to move to Australia, where my passion for technology eventually led me into the academic world. First, I pursued a Master of Engineering in telecommunications. Then, driven by a desire to contribute to human knowledge and my long-term dream, I also sought a doctorate in telecommunications engineering.

Over all these years of study and work in the telecommunications sector, when meeting someone or at many social events, conversations are usually directed to what we do or have studied. With a few exceptions, mainly those working or studying in a similar sector, almost no one is genuinely interested in what I have to say. People recognize the nature and the importance of telecommunications and technology in their daily lives, but these topics don't usually captivate someone outside these fields.

During my four-year doctorate, and similar to what I experience when I describe my day job, very few people asked me about the nature of my research or expressed a profound interest in the topics I was working on. Trust me, it was not because of a lack of passion on my part when sharing my experiences.

In contrast, as soon as the conversation shifts towards my interest in astronomy, that completely changes. Ever since I was a kid and began exploring these subjects, I noticed that all kinds of people were eager to know more. Just set up a telescope in a park or on a city street at night, and you will see how people approach to peek and start asking questions about what they have or have not seen.

One of my more vivid memories from my youth is talking to a group of my cousin's high school friends. There we were, four 14-year-old kids discussing cosmology and the universe's fate on a Wednesday afternoon after school.

My mid-life crisis did not involve buying a Ferrari (not that I had the money to do it anyway) or making rush decisions (we did move to Queensland after 12 years in New South Wales, though). Instead, it was a realization that I needed to return to my dream of formally pursuing astronomy. That's when I enrolled in a Master of Science in astrophysics.

During my astronomy degree, people were intrigued by my studies and would ask me questions about the universe, planets, and the possibility of extraterrestrial life. Their enthusiasm was evident for talking to someone who was getting formally educated in these topics. Only my wife's subtle cues (or a nudge from her elbow when she was nearby) prevented me from monopolizing the conversation. Especially since I've shared these stories with her countless times during all our years together.

Such episodes made me realize that it was time to pursue my other long-term dream: writing a book; and here we are. This book is a dream come true and an attempt to explain in simple terms many of the wonders that captivate me and, based on multiple conversations I've had over these

Preface

years, a lot of other people. It is also a tribute to all the researchers scientists who have devoted and continue to devote their lives to pursue their dreams and passions and those who remain curious.

For all of you out there who share this passion for the universe, this book is dedicated to you.

Introduction

"Space: the final frontier."

— Captain James T. Kirk, Star Trek, original series (1966)

Out of everything else in the world, what intrigues me the most, and I believe, to many, is the possibility of the existence of life beyond Earth. For this reason, the possibility of extraterrestrial life is one of the main subjects of this book. However, life as we know it requires a planet, and until not too long ago, the search for such a life was strictly limited to the planets within the solar system. However, everything changed in 1992. It was this year that the first planet orbiting a star other than the Sun was detected, and therefore, the existence of other planetary systems was confirmed. Since then, the discovery of a new extrasolar planet, or exoplanets as they are known, continues to be announced almost every day. In April 2024, NASA highlighted that more than 5,600 exoplanets have been detected, hence the title of this book: *A Story of More Than*

Introduction

5000 Worlds. This growing number only increases our chances of finding life beyond Earth.

Planets are, however, a byproduct of the star formation process. In this book, we will explore the formation of stars and planets, examine different techniques to detect, characterize, and classify exoplanets, and discuss the significance of the search for extraterrestrial intelligence (SETI). We will discuss topics such as the Drake Equation, the Fermi Paradox, and the latest advancements in life detection.

Before we begin, it's important to clarify the meaning of "hypothesis" and "theory" in science, as we will use these terms frequently throughout the book.

In science, contrary to the language we use every day, the word "theory" refers to an explanation of a natural phenomenon. It is an explanation tested and confirmed through observation and experimentation. This does not mean that the explanation won't change in the future, but it means that it fits well with our current understanding and observing capabilities. The most important aspect is that scientific theories can make testable predictions of the phenomena they explain. To achieve this, a scientific theory must be structured so that it can potentially be proven false by an experiment or an observation. Scientists refer to this property as the *falsifiability* of the theory. Notable examples are the theory of the big bang, the theory of relativity, and the theory of evolution.

On the other hand, the meaning of the word "hypothesis" in science is equivalent to the meaning of the word "theory" in the everyday language (confusingly enough?). In science, a hypothesis is an attempt to explain a phenomenon. However, a hypothesis is usually a starting point for investigation. Traditionally, it is based on observations, previous knowledge, and existing theories. It is a speculative educated guess that yet needs to be tested. We will explore many theories and hypotheses and will hopefully get you, the reader, interested in exploring these topics even further.

Two additional important definitions to consider, which we will also use often throughout the book, have to do with how astronomers express the distances between objects in space. The universe is so vast that the

Introduction

regular meter and kilometer units we use every day fall short when describing such distances. The first of those units is the light-year, defined as the distance that light travels in one year. The speed of light is the maximum speed in the universe, and it is about 300,000 kilometers per second. Therefore, a light-year is equivalent to approximately 9.46 trillion kilometers (a trillion is a 1 followed by twelve zeros). It is also common for astronomers to express the distance that light travels in one hour, as a light-hour, in a minute, as a light-minute, or in a second, as a light-second. For instance, the Sun is located at 8 light-minutes from Earth, meaning that light requires 8 minutes to travel the distance between the Sun and Earth. On the other hand, the Moon is much closer to Earth, at a distance of approximately 1.28 light-seconds. This is important because the signals we use to communicate travel at the speed of light, which is a reminder that communication is not instantaneous, despite of what we see in a lot of sci-fi movies.

The second unit of distance is the parsec, whose definition requires a couple of elements that will be introduced in Chapter 4, Exoplanets Detection Methods – Second Part. For now, let's just indicate that a parsec equates to 3.26 light-years or approximately 30.86 trillion kilometers.

This foundational knowledge allows us to delve deeper into astronomical concepts. This book, in addition to acting as an engaging read for anyone interested in these subjects and given the number of topics included and the breadth of the discussions, could also be used as a complementary textbook for students at both high school and non-science bachelor levels, in introductory or fundamental astronomy courses.

For quick reference, I have included a *Too Long; Didn't Read* (*TL;DR*) section at the beginning of each chapter. This *TL;DR* section is designed to offer readers a summary of the key points discussed in the chapter.

Let's dive in!

Chapter 1
Stars

TL;DR

On Earth, the Sun, its parent star, provides the necessary energy for life to exist. But the Sun is only one of many stars in the universe.

Our current understanding indicates that stars form when a huge cloud of gas and dust in space gets pulled together by gravity. These clouds are mostly hydrogen gas, the most common element in the universe, which forms the building blocks of new stars. From a certain threshold of initial mass, a star begins its main life stage by turning hydrogen into helium through fusion. Once the star has burnt all its hydrogen, the outer layers of the star begin to expand. Low and medium-mass stars turn into red giants, whereas high-mass stars become red supergiants.

During the red giant stage, stars grow so much that any nearby planet orbiting the star is engulfed. After stars have finished expanding, low and medium-mass stars will expel their material into the universe in the form of a planetary nebula. At the center of the planetary nebula, a white dwarf remains. White dwarfs irradiate all its energy over eons, eventually turning into black dwarfs. Unfortunately, we have not been able to observe black dwarfs as there has not been enough time in the universe's history for white dwarfs to radiate away all their energy.

In contrast, high-mass stars will explode after their red supergiant phase. These explosions are called *supernovae*, and they are amongst the brightest events in the universe. After the explosion, the cores of these stars will contract even more due to their still remaining high mass. Astronomers estimate that for stars with initial masses between 10 and 25 times the mass of the Sun, the core will contract until a stable neutron star forms. Stars with initial masses above 25 times the mass of the Sun will contract even more, leaving behind a black hole. Figure 1.1 presents a nice and concise picture of the lifecycle of a star.

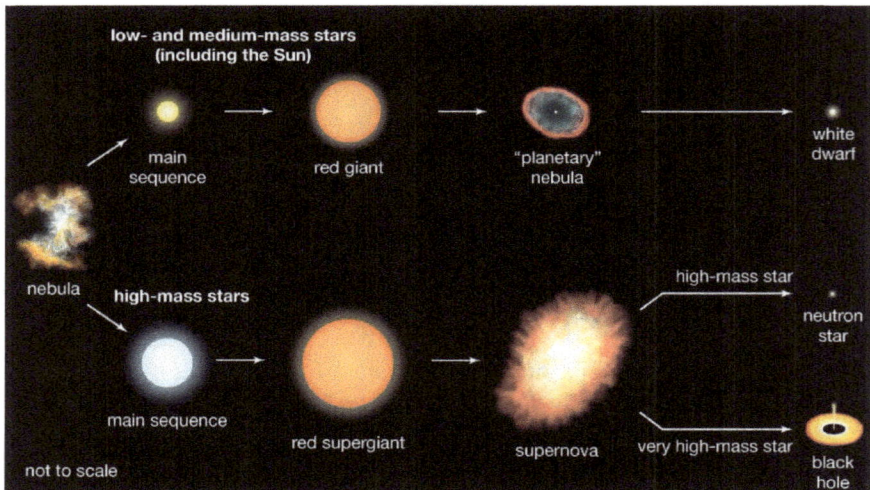

Fig. 1.1: The lifecycle of stars. Stars' fate is tied to their initial masses.
Credit: 2012 Encyclopedia Britannica, Inc.

Our Parent Star - The Sun

Our destiny has always been linked to the stars. Users of the Gregorian calendar have selected the last night of December and the first day of January as a marker for when our planet completes a full orbit around its parent star, the Sun. We cry, laugh, and hug our loved ones while we all yell "Happy New Year!".

At a more personal level, parents may mark in calendars the day when we are born. Then, every time the Earth goes around the Sun one full orbit, we celebrate our birthday and continue to do so for the rest of our lives.

In many ways, the Sun is considered a very ordinary star. It has been shinning for over 4 billion years and it is comprised of 71% hydrogen, 27% helium and 2% of all other elements combined.[1] We will learn that such a distribution of hydrogen and helium is not unusual among other stars.

With a radius of 695,700 kilometers, the Sun is so huge that it could fit 1.3 million Earths. It is also extremely heavy with a mass of nearly two nonillion (that is a two followed by 30 zeros) kilograms. Our star contains 99.8 percent of the mass of the entire solar system.

The solar system is part of the Milky Way, our local galaxy. Galaxies are systems of stars, stellar remnants, interstellar gas, dust and dark matter, all this bound together by gravity. The Milky Way has a diameter of 100,000 light-years and is home to approximately 100 billion stars (one billion is a 1 followed by nine zeros). The Sun is located about 25,000 light years away from the galaxy's center.[2,3]

The Sun is crucial to life on Earth as we know it. Water can exist in liquid form on Earth's surface in part because of its distance from the Sun. That is, we say that the Earth orbits the Sun in its habitable zone (HZ), also known as the Goldilocks zone (we will talk more about this throughout the book). Sun's habitable zone has been estimated to be between 0.95 to 1.15 astronomical units[4] (an astronomical unit (AU) is the average distance at which the Earth is from the Sun and equates to 150 million kilometers).

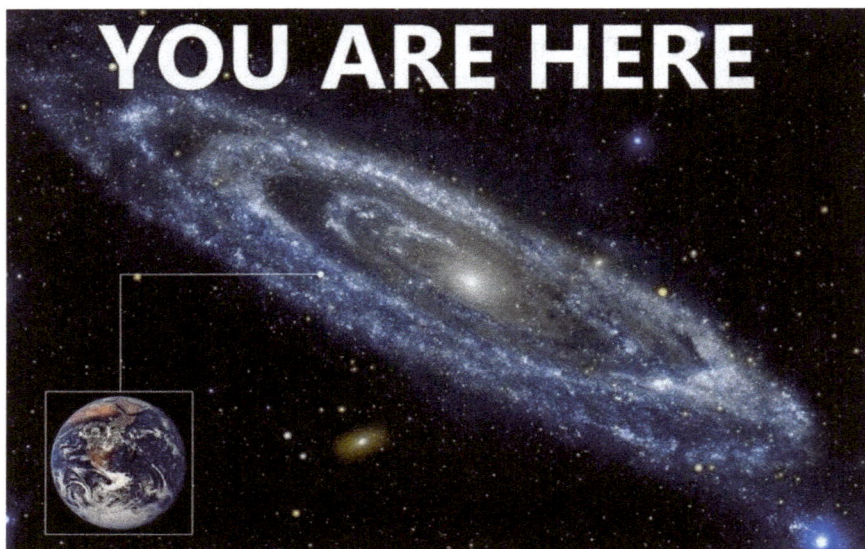

Fig. 1.2 The Earth orbits around the Sun, one of the 100 billion stars in the Milky Way. Credit: The Universal Story by Aidan Zellner.

Water will remain liquid on a planet's surface, if the planet is at the right distance from its home star, i.e., is located within the stars' HZ, and possesses the right atmospheric pressure conditions and atmosphere characteristics.

On Earth, the Sun provides the required energy for essentially all organisms to survive—with the notable exception of life found on the bottoms of oceans. Sunlight is also the key for multiple chemical processes to take place. The most known chemical process is photosynthesis, which is the process that occurs in plants whereby energy obtained from light is used to convert carbon dioxide and water into glucose and oxygen.

For us humans, the Sun has been, and will always be—for most of us at least— in the sky providing its energy and majestic presence.

Many ancient cultures, often considered the first astronomers,[5] acknowledged the Sun's role as the provider of life on Earth. For instance, the large yellow circle in the official flag of the Australian aboriginal people represents the Sun, the giver of life and protector.

4

Fig. 1.3 The official Australian aboriginal flag. The black symbolizes Aboriginal people. The yellow represents the Sun, the constant re-newer of life. Red depicts the earth and peoples' relationship to the land, and also represents ochre, which is used by Aboriginal people in ceremonies.

Any other potential life out there will require the right conditions from their host star to thrive as well. But how did the Sun, and all the stars, form?

The Lifecycle of Stars

We will begin our journey by discussing what we currently know about the lifecycle of stars. But how do we know what we know? We will start by answering that question upfront. Lifecycles of many stars last billions of years and the ones that die faster, massive stars, live for millions of years. This means that humans are not able to witness the whole process from the beginning to the end. The universe is full of examples like these where the timeframes of the phenomena at hand is not within a person's lifetime, and therefore, astronomy is full of analogies. This is one of the aspects that attracts me the most about this discipline: the lengths to which astronomers and science communicators must go to explain difficult concepts in simple terms.

Let's assume that you want to study human beings from birth to death. We could choose only one individual and document that person's life throughout the years. Of course, this is not necessarily practical or achievable. Life expectancy currently sits at around 80 years. That would be a lot of YouTube videos or our own version of *The Truman Show*, a very good 90's movie starred by Jim Carrey. However, if we observe different human beings at different ages at the same time, we can come up with a pretty good picture of how humans behave and what they look like throughout their lives. Imagine watching a game of the most popular football team in Colombia, America de Cali (according to me and millions of other knowledgeable football people), where fans of all ages come together.

Fig. 1.4 Emblem of the football team America de Cali, one of the most awarded teams in Colombia. Founded in 1927, it is considered by many, and by far, the best football team in the country.

When you get to the stadium, with a set of binoculars to observe the game even better, you notice that opposite to where you are sitting, a mother has taken her new-born child to the stadium. When you look to your right you realize that there are some kids close to you. These kids are as young as two, four, six, and 10 years old. You keep looking and notice some older kids, maybe 16 by judging for their incipient beards that start to show in their faces and their 'who cares' attitude. You then find some adults; some of them are still quite young, perhaps, early twenties, but you also find some older people. A couple in their thirties, a group of friends in their forties and fifties. You then keep looking and find people in their golden ages. People in their sixties, seventies, and even eighties. From this quick exercise you conclude two things: First, this football team attracts people from all ages, and second, that you have seen the progress of a human being through all the stages from the moment they are babies until an advanced age. You are now ready to at least identify or even classify the different stages human beings go throughout their lives.

With this simple but useful analogy, you can get an idea of how astronomers study stars at different stages of their lifecycle and construct an accurate depiction of their lives. Astronomers observe stars that are literally just getting born, others that are just minding their own business and living their lives, and others that are dying. They even observe their remnants after their deaths. We humans want to understand things. Especially if those things relate to us.

Star Formation

The star formation process is still a very active research topic. However, something astronomers have a lot of evidence for is that a star's life begins with *clouds*. More specifically, *giant molecular clouds (GMC)*, or stars nursery regions as they are also called. We can observe one of those star-forming regions in a beautiful image released by NASA in July 2022.

Fig. 1.5 Star-forming region NGC 3324 in the Carina Nebula as captured by James Webb Space Telescope (JWST). Credit: NASA, ESA, CSA, and STScI via AP.

Star formation is a very complex process that involves understanding complicated gas dynamics and magnetic field interactions. The *Theory of Star Formation*[6] has been and continues to be the topic of many publications, including books and research articles.

Giant molecular clouds (GMCs) are comprised of gas and dust in regions of the interstellar medium. Astronomers refer to everything situated between stars as *interstellar medium* (ISM). The gas component of the ISM predominantly consists of hydrogen, the most abundant and lightest element in the universe. Hydrogen accounts for about 75 percent of the matter made out of atoms and was created through a process known as *big bang nucleosynthesis*[7] during the very first stages of the Big Bang.[*]

The dust in the ISM is literally that: dust. In his very entertaining and provocative book "Losing the Nobel Prize",[8] author Brian Keating calls dust "the astronomer's antagonist". Such a description is not without reason. Dust obscures the light of objects that astronomers want to observe and analyze, causing uncertainty about the light brightness of a given source. Astronomers call *extinction* to the reduction in brightness due to absorption and scattering of dust particles. Even more, extinction also affects different colors of light differently, making it harder for astronomers to reconstruct the true color—spectrum— of a source. An extinction value needs to be included in calculations and models to accurately establish the distances, sizes, and temperatures of celestial objects. For instance, the importance of dust in accurately measuring the Hubble constant has been largely discussed.[9] The Hubble constant, named after Edwin Hubble, measures how fast the universe is expanding. One method that astronomers use to measure the Hubble constant consists of capturing light from very distant objects such as supernovae and calculating the distance at which they are from us, as well as the speeds at which they are moving away from us. Another method involves measurements of the Cosmic Microwave Background (CMB), the initial thermal footprint left by the Big Bang. The discrepancy in the results obtained using these two methods has been dramatically described as "the crisis in cosmology" or the "Hubble tension".[10]

Molecular clouds do not have a uniform density. Instead, some regions are denser than others. These regions, or small regions within the larger

---

[*] Helium and a small amount of Lithium were also created during the first few minutes after the Big Bang. Heavier elements were and continue to be created during the life cycles of stars.

region, start experiencing a disequilibrium between the force of gravity and the thermal pressure within the gas resulting in the *fragmentation* of the cloud into smaller clouds. Gravitational instabilities cause these small clouds to become self-gravitating cores that start to collapse under their own weight.

Due to gravitational instabilities, these collapses can unfold over millions of years. However, a collapse can also be precipitated by a shockwave from a highly energetic explosion nearby, such as a supernova. We will explore supernovae in greater detail later in this chapter.

Regardless of the mechanism, the collapse of a molecular cloud highly depends on the initial temperature, average density of the cloud, and the cloud's chemical composition. For instance, for a hydrogen cloud with an initial temperature of 10 Kelvin (-263 degrees Celsius), the required mass for the collapse to occur is approximately eight times the mass of the Sun.[11] This means that only if this mass is present will the collapse of the cloud start to occur.

Initial conditions might cause the cloud to expand instead of collapse. 10 Kelvin is really cold. The Kelvin scale is based on the concept of absolute zero. Temperature is a measure of the energy that a particle in motion possesses. In physics, this energy is known as *thermal energy*. At absolute zero, there is literally no movement of particles. This is the lowest possible temperature that can exist. Absolute zero corresponds to -273 degrees in the Celsius scale.

Now, you may or may not remember the physics class in high school when the concept of *conservation of angular momentum* is introduced. The angular momentum is a fundamental concept in physics; it is a measure of the motion of an object rotating around a fixed axis. To increase the angular momentum of an object, you can increase its rotational velocity (how fast the object rotates), or you can also either increase the mass of the rotating object or the distance at what the object is from the axis of rotation. For an isolated system, i.e., a closed system that is not under the influence of external forces, the total amount of angular moment won't change over time.

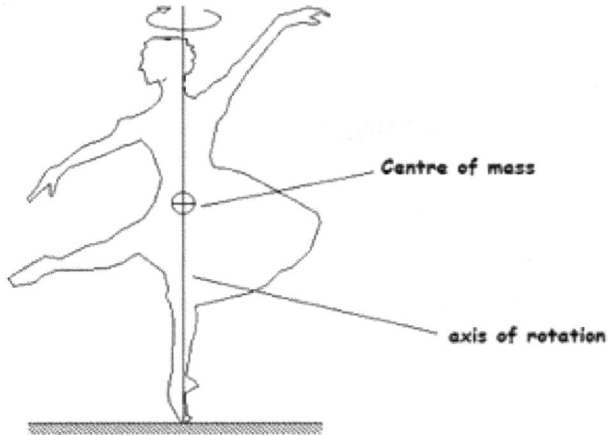

Fig. 1.6 Dancers will take advantage of the conservation of the angular momentum to control their rotational speed. Credit: The physics of ballet.

The typical example is a spinning ballet dancer. The ballet dancer is considered a closed system, assuming no other dancer exerts force over them. When a dancer wants to spin slower, they just need to stretch their arms, effectively increasing the distance from the axis of rotation to their own center. As the total angular momentum needs to be constant, rotational speed needs to decrease. On the contrary, if they want to spin faster, they fold their arms, reducing the distance between their own center and the axis of rotation, resulting in higher rotational speed.[12]

This is precisely what happens in a molecular cloud. As the cloud starts to collapse, its rotational speed increases. Next time someone asks you what molecular clouds in space have in common with spin dancers, you know what to say.

The increase in speed causes the cloud to flatten, effectively becoming a flat rotating disk, known as *protostellar disk*, with a collapsing protostar at its center. The formation of the disk is a way to conserve the angular momentum that's going toward the star at the center. We can observe the same process in the art of pizza tossing. Cooks take advantage of forces acting outwards on a body moving around a center. This *centrifugal* force pushes the outer edges of the dough outward, stretching it and causing it to expand in diameter and flatten.

Fig. 1.7 The flattening of pizza dough. The resulting flatness is the result of centrifugal forces acting outwards on a rotating object. Credit: the public Internet

At this stage, the temperature of the cloud and the disk continues to rise due to the effects of gravity and pressure within the gas. Temperature increases when gravitational potential energy is transformed into thermal energy. Also, as the cloud becomes denser, given that the same mass is occupying less physical space, particles within the cloud start to collide more frequently. These collisions cause thermal energy to increase. At the same time, the core is becoming extremely hot, increasing outwards pressure within the gas. Gravity is always trying to collapse the matter in the core, whereas pressure is trying to push it outwards.

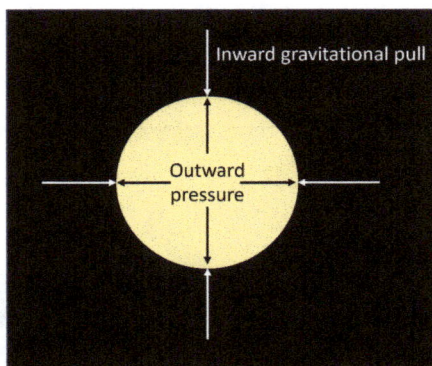

Fig. 1.8 The constant battle between gravity and pressure in a star. Credit: image by the author.

The giant molecular cloud has now been transformed into a *protostar*—an embryo star— surrounded by a *protodisk*. However, it is still not hot enough for atoms of hydrogen to *fuse* and form one single atom of helium. Such a process is called *fusion*, and when this happens, the embryo becomes officially a star and enters what is referred to as *the main sequence*.

Pre-main sequence Stars

You may be getting impatient (I don't blame you) and have started asking yourself: when will we start talking about actual stars? Those that produce their own light? Please bear with me as we discuss one last stage before a star makes it to the main sequence. There's an intermediate stage between a protostar and a main sequence star, known as a *T Tauri star* if the protostar at the center of the protodisk has a mass less than two times the mass of the Sun or as a *Herbig Ae/Be star* if the protostar's mass is between about 2 and 10 times the mass of the Sun.[13] Despite these "baby stars," or *pre-main sequence stars*, as they are formally known, not yet undergoing fusion in their cores, they possess leftover heat from their formation and may be capable of accreting material from the surrounding disk, contributing to their luminosity.

Think of these baby stars as young stars that just entered the stellar kindergarten. Similar to regular kindergarten kids, they make a lot of noise and are quite active. These stars exhibit frequent flares, eruptions, and strong stellar winds. They also exhibit high variability in their brightness and have strong magnetic fields. As material surrounding the original molecular cloud continues to fall at the center, temperature rises.

Not all molecular clouds form a *T-Tauri Star* or a *Herbig Ae/Be star*. Sometimes, these clouds do collapse, but the initial mass is not large enough, or the temperature does not get high enough, and nuclear reactions do not occur. Such objects are known as Brown Dwarfs, and we will talk a little bit more about them in a moment.

Nuclear Fusion

Okay, we are finally here. For a typical pre-main sequence (PMS) star, efficient hydrogen fusion starts when the core reaches around 10 million Kelvin (roughly 10 million degrees Celsius too). Without going too deep, we remember from high school the periodic table and how each element had a little number on top of its symbol. This number is called the atomic number and represents the number of protons in the nucleus of a single atom of the element. Hydrogen is the lightest element with only one proton; hence, its atomic number is 1.

Fig. 1.9 Hydrogen is the simplest and most abundant element in the universe.

When fusion occurs, through a series of reactions called the proton-proton chain, four hydrogen nuclei fuse to form a single helium nucleus.

13

Even though some mass is converted to energy, the resulting helium atom still has two protons, which is its atomic number.

Fig. 1.10 Helium is the second simplest and most abundant element in the universe.

The point in time when a star achieves hydrogen fusion is called *The Zero-Age main sequence* (ZAMS). Therefore, the *main sequence* represents a period when stable hydrogen fusion in a star's core occurs. The main sequence is depicted as an almost straight diagonal line in what is called a *Hertzsprung-Russell diagram*, or *H-R diagram*[14] for short. The H-R diagram is an extremely important and useful tool in astronomy. The diagram, presented in Figure 1.11, plots the brightness, or luminosity in solar units, of close-by stars against their temperature in the Kelvin scale.

Like other terms in astronomy, where names don't necessarily correlate to the actual nature of what is being described (*planetary nebula*, for instance, has nothing to do with planets), the main sequence is not a sequence at all. A star does not begin as a given type of star and then evolves into another type of star in the sequence. Stars enter the main sequence at a certain point depending on the resulting mass at the end of the formation process.

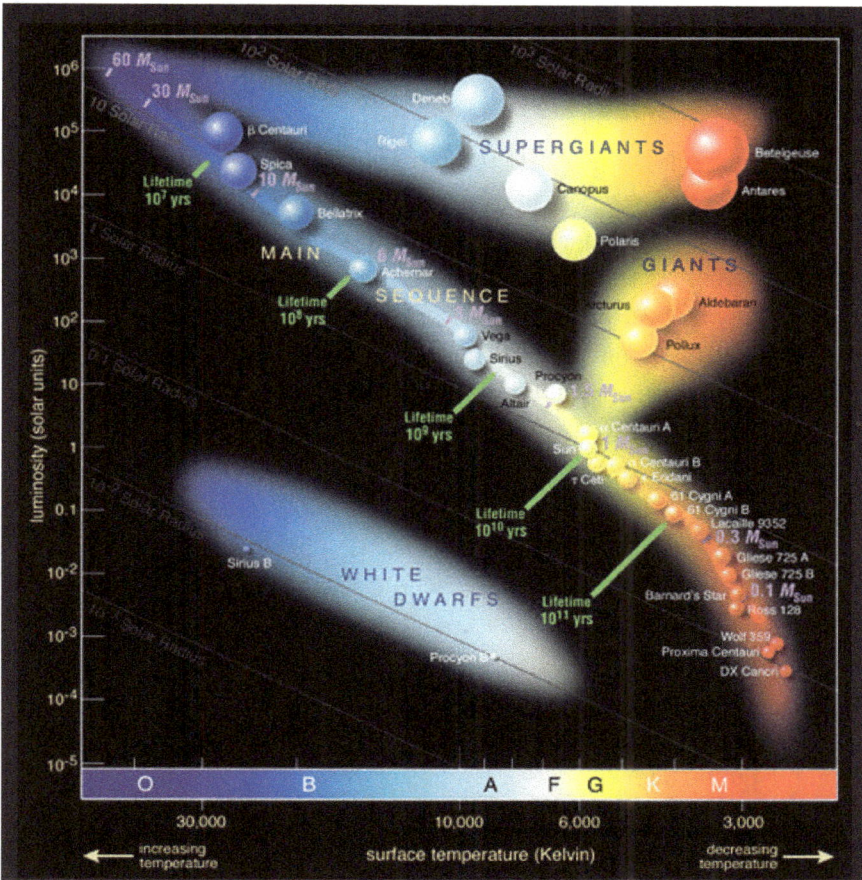

Fig. 1.11 The Hertzsprung-Russell diagram, or H-R diagram for short.
Credit: The cosmic perspective (p. 832).

The mass of the resulting star is directly related to the mass of the original molecular cloud. Molecular clouds don't remain intact during their existence and will undergo fragmentation due to gravitationally instabilities. A massive molecular cloud will produce more massive fragments, which will result in more massive stars. The initial mass of the cloud also determines how much time the cloud takes to collapse. Or, in other words, how long a star takes to initiate nuclear fusion. The larger the mass, the shorter the contraction time. For instance, a molecular cloud with an initial mass of 15 times the mass of the Sun may collapse in "only" 60,000 years. On the other hand, a cloud with an initial mass equivalent to half the mass of the Sun's would take around 150 million years to contract.[15]

A star's mass is the most important determinant of its temperature; there is an intrinsic relationship between mass and how much energy stars radiate. More massive stars are more luminous as nuclear fusion is more efficient. Remember that nuclear fusion directly results from the force of gravity winning the battle against the gas pressure to expand the star. It is worth noting that astronomers distinguish between luminosity and brightness. Luminosity being an intrinsic measurement of the amount of energy emitted by an object, and brightness being affected by the object's distance from us.

The definition of 'Metals' in Astronomy

Astronomers are extremely busy people. They must divide their time doing research, teaching courses, supervising students, elaborating and contributing to scientific papers, grants, and telescope observing proposals. In addition, a lot of them feel an obligation to the taxpayers to perform public outreach as, for most of them, their wages come directly from the public budget.

Public outreach benefits everyone. Many scientists feel quite passionate about it as it helps them connect with the public. Outreach also increases public awareness and understanding of specific issues, such as climate change, to provide just one example. Increasing public understanding of fundamental science can lead to more informed choices and potentially lead to more funds put into someone's research.

With all these obligations, astronomers can't waste their time on trivial things such as naming chemical elements one by one. This is why, in astronomy, the term "metal" refers to any element that is not hydrogen or helium.

Given that hydrogen and helium are by far the most abundant elements in the universe, the use of the term "metal" helps simplify the language and communication.

Despite Figure 1.12 being intended as a joke, it is not far from reality. The image shows the periodic table according to astronomers.

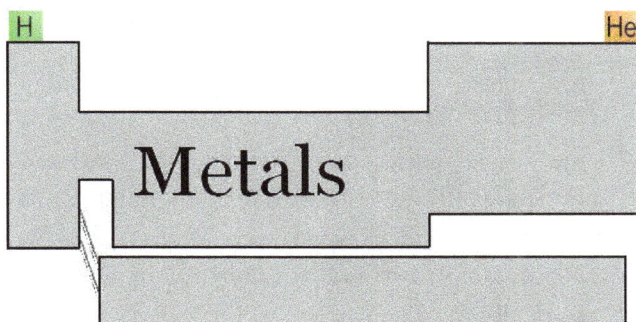

Fig. 1.12 Astronomers' definition of a metal is anything heavier than hydrogen and helium. Credit: Taken from reddit.

Classification of Stars

Stellar Population

Stars can be classified according to their chemical composition. This classification is useful because it provides astronomers with an indication of when a star formed relative to the universe's age.

We mentioned that stars fuse lighter elements into heavier elements in their cores, provided the temperatures and pressures are sufficient to do so. Hydrogen can be fused into helium, helium into carbon (as we will see later in this chapter), and so on. However, in the beginning, the universe was an environment where only hydrogen and helium were present. Consequently, the first stars—first-generation stars— formed during this time condensed out only of hydrogen and helium. In other words, these first-generation stars reflect a complete absence of metal elements in their composition. Remember that astronomers consider metals everything that is not hydrogen and helium. These primordial stars are referred to as Population-III stars.

Population-III stars

Logic dictates that astronomers should have named these primordial stars as Population-I stars (since I comes before III), but one more time, naming things logically rather than sticking to historical reasons is not really something the field of astronomy is known for.

Population-III stars have been purely hypothetical until now, as their existence has not been confirmed. These stars are thought to have formed between one million years and 1 billion years after the Big Bang.[16] They could be very massive, up to several hundreds of times the mass of the Sun, due to inefficient fragmentation of the original giant molecular clouds in the early universe. Due to their large masses, these stars are also likely to have been very hot and capable of emitting high-energy photons sufficient to double ionize helium (the removal of two electrons from helium atoms), which hints at what type of signatures astronomers need to look for in the spectra of potential candidates. The discovery of such objects requires instruments capable of looking far back enough in time, as close to the Big Bang as possible. Enter the James Webb Space Telescope (JWST), launched on December 25, 2021, and considered the most advanced telescope at present. With a total cost of around 10 billion dollars, this instrument is helping astronomers to confirm and sometimes redefine what we know about the universe.

In 2023, astronomers using the JWST indicated that they may have found evidence of these elusive Population-III stars. In their article,[17] the researchers report observed sources of double ionized helium originating between 150 million to 1 billion years after the Big Bang. Although active galactic nuclei (which we will discuss in a moment) and nearby black holes could also emit such radiations, their analysis has ruled out these sources. The authors acknowledge that more observations are necessary to confirm such findings.

Population II Stars

Population II stars are believed to have originated around one to two billion years after the Big Bang. They are the oldest stars astronomers can currently observe. Unlike Population III stars, Population II stars did have access to heavier elements or metals, spread through the inter-stellar medium by supernovae or high energetic processes resulting from the deaths of Population III stars. However, these heavier elements were not available in large quantities at the time when Population II stars formed. For this reason, the composition of these stars is *metal-poor*; in other words, their abundance of metals, referred to as the *metallicity* of the star, is low. Population II stars are less massive, with masses between 8 and 40 times the mass of the Sun. This is mostly because the existence of metals allowed further cooling and smaller fragmentation of the giant molecular clouds they come from,[18] as compared to the giant molecular clouds from where Population III stars originated from. It is estimated that the abundance of metals in Population II stars is only one-tenth or one-thousandth of that of the Sun.[19] One of the brightest, nearby, metal-poor Population II stars is HD 122563. This red giant star is located at approximately 1,050 light-years from Earth and has an estimated age of 12.6 billion years.[20] The spectral analysis of HD 122563 revealed that the star has one thousandth metal abundance relative to hydrogen compared to the Sun.[21]

Population I Stars

Contrary to Population II stars, Population I stars are younger and more metal-rich. These larger metallicities result from an environment where heavier elements, the result of Supernovae explosions and planetary nebulae from Population II stars deaths, enriched the seed material from which these younger stars formed. Our star, the Sun, with "only" 5 billion years of age, is categorized as a Population I star.

Population I stars can be as old as 10 billion years old or still forming today. In these stars, heavy elements account for between one and four percent of their total stellar mass.[22]

The metallicity of stars can play a pivotal role in the emergence of complex life. Researchers have shown how the metallicity of a star is connected to the ability of the orbiting planets to develop an ozone layer.[23] The lower the metallicity, the more life-friendly the star. Similar to what happens on Earth, a surrounding ozone layer could protect the cells of potential life on a planet's surface from the host star's ultraviolet radiation.

Temperature as Stellar Classification

Stars are also classified according to their apparent temperature. Despite the classification being quite useful and simple (from hottest to coolest), the names of the categories, which are just letters, are non-intuitive nor user-friendly. The scheme results from multiple attempts by different people to develop a useful framework over the years.

The temperature sequence "O B A F G K M" or spectral classification, helps astronomers to quickly identify and categorize stars.

Light is made up of waves that travel through space. These *light waves* have important characteristics that allow scientists to classify them. One of these characteristics is the concept of *wavelength*. The wavelength of a light wave is defined as the distance measured in meters between two consecutive 'peaks' (high points) or two consecutive 'valleys' (low points) of a wave.[24] Due to how small such wavelengths could be in meters, scientists use the *nanometer* unit. One nanometer is one billionth of a meter (1/1,000,000,000) and is expressed as *nm*.

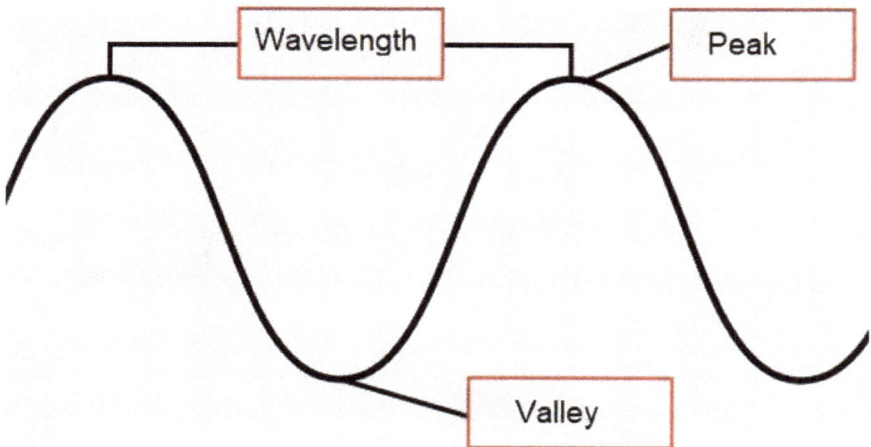

Fig. 1.13 The wavelength is the distance measured in meters between two consecutive peaks or two consecutive valleys of a light wave. Credit: modification of work by lumen.

When light traverses a prism, we can observe its different components. Astronomers call this a light spectrum, which refers to the range of electromagnetic radiation wavelengths that we can see (visible spectrum) and those that we cannot see (radio, microwave, infrared, ultraviolet, X-rays, gamma ray).

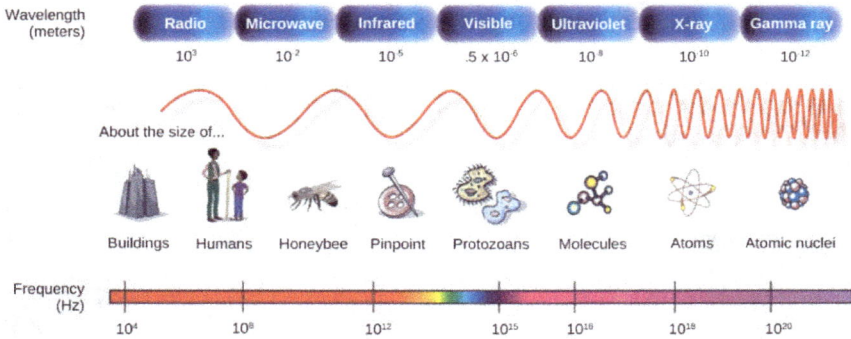

Fig. 1.14 The electromagnetic spectrum. Visible light is in the range between 380 nm and 740 nm. Credit: lumenlearning.com, modification of work by NASA.

The visible spectrum is in the range between 380 nm and 740 nm and is typically divided into seven main colors: red, orange, yellow, green, blue, indigo, and violet. Each color corresponds to a different wavelength of light, with red having the longest wavelength and violet the shortest.[25]

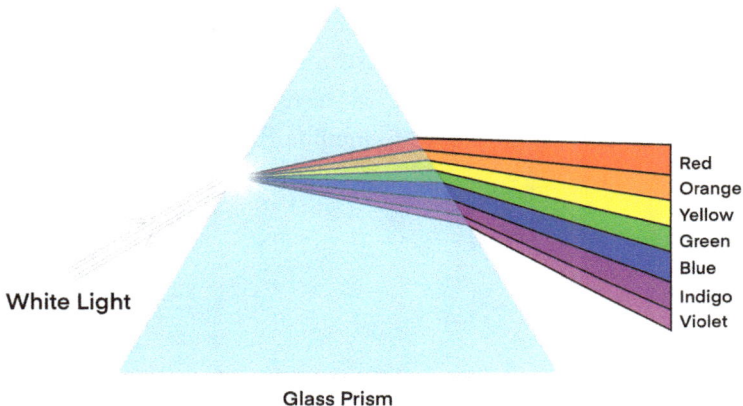

Fig. 1.15 Visible light is broken apart after passing through a prism. Credit: Florida State University.

There is a clear relationship between temperature and color (wavelength). For objects that emit thermal radiation, there is a peak wavelength at which they emit the brightest. The higher the temperature, the shortest the peak wavelength. The "O B A F G K M" sequence starts with the hottest, blue 'O' stars and ends with the coolest red 'M' stars (also known as red dwarfs).[26] The Sun is a 'G' type yellow star with a temperature of approximately 6,000 Kelvin (5,726.85 degrees Celsius).

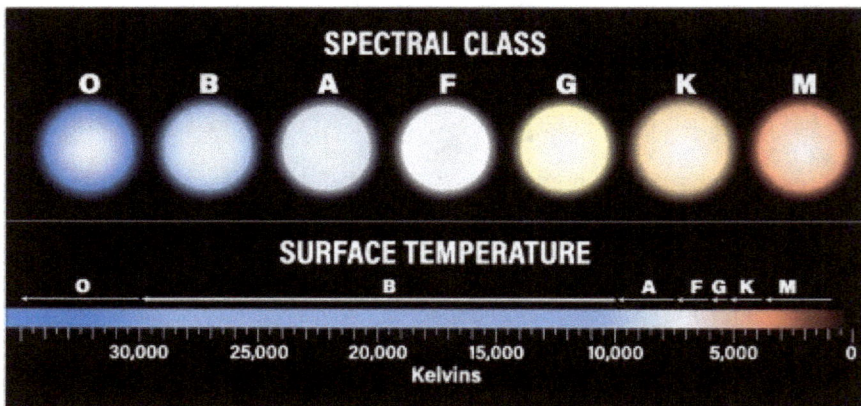

Fig. 1.16 Stars are classified based on their colors (spectral class) which have a direct correlation with their temperatures. Credit: Astronomy Magazine.

Astronomy students have come up with interesting phrases to remember the classification. A common one is "Oh Be A Fine Girl/Guy Kiss me". However, more politically correct acronyms such as "Oh Boy, An F Grade Kills Me" or "Only Boring, Astronomers Find Gratitude Knowing Mnemonics" have been suggested. More recently, three new categories, L, T, and Y, have been introduced to include *Brown Dwarfs*.

Brown Dwarfs

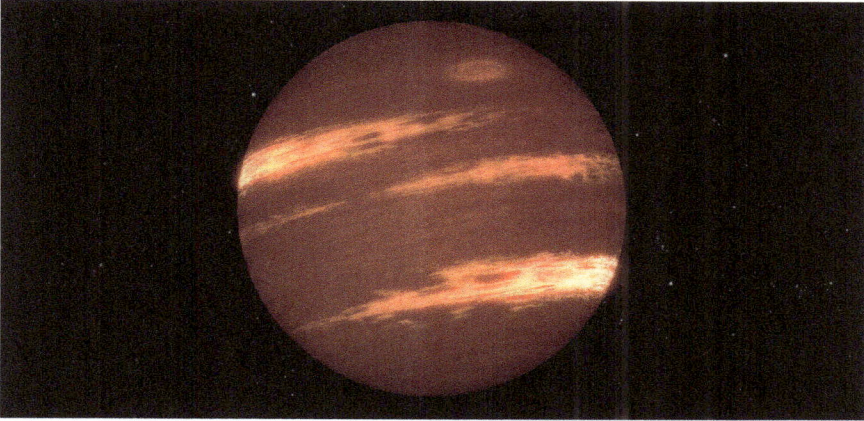

Fig. 1.17 An artist's depiction of a Brown Dwarf. Credit: NASA, ESA, and JPL-Caltech.

The term *Brown Dwarfs*, which refers to stars that belong to the 'L', 'T', and 'Y' categories, was introduced by the founder of the SETI institute and SETI super star Jill Tarter.[27] Brown Dwarfs are failed stars or stars not massive enough to be able to fuse hydrogen. Their masses vary between 13 to 80 times the mass of Jupiter. 'L' class stars have temperatures between 1,300 Kelvin (1,026 degrees Celsius) and 2,200 K (1,927 degrees Celsius), whereas 'T' stars have temperatures between 800 Kelvin (527 degrees Celsius) and 1,300 Kelvin (1,026 degrees Celsius). 'Y' dwarfs are the coldest stars-like objects in the universe (that we are aware of). These objects have temperatures below 800 Kelvin (527 degrees Celsius). There have been even observations of Brown Dwarfs that exhibit temperatures similar to the average room temperature of 27 degrees Celsius (300 Kelvin).[28, 29]

Red Giants

By knowing a star's temperature, astronomers can infer their mass. This is important because a star life's expectancy is determined by its mass. Given that massive stars are more efficient at burning up their fuel supply, their lives are shorter. Astronomers usually describe massive stars as the ones that "live fast and die young", making reference to the hit song "Live Fast, Love Hard, Die Young" by Faron Young in 1955; and this is true. Very massive stars can exhaust their fuel, or in other words, leave the main sequence, in as little as a few million years. On the other hand, very small stars can take longer than the age of the universe to consume all of their hydrogen. A medium mass star like the Sun, takes around 10 billion years.

The mass of a star also determines what it becomes after it has left the main sequence. Low and medium-mass stars, like the Sun, become *Red Giants* once they have exhausted their hydrogen supply, fusing to helium at their cores. Red giants are between 100 and 800 times the size of the Sun. Similar to what happened in the beginning with the original molecular cloud, gravity starts winning the battle against gas pressure. These stars have atmospheres that grow large and consequently have larger radii. At this stage, the core, which is now comprised almost entirely of helium, starts to shrink while the outer layers start to expand. For comparison, Figure 1.18 shows how the Sun will grow in size, "devouring" Mercury and Venus and putting the Earth in a pretty awkward position where life is not supported anymore.[30]

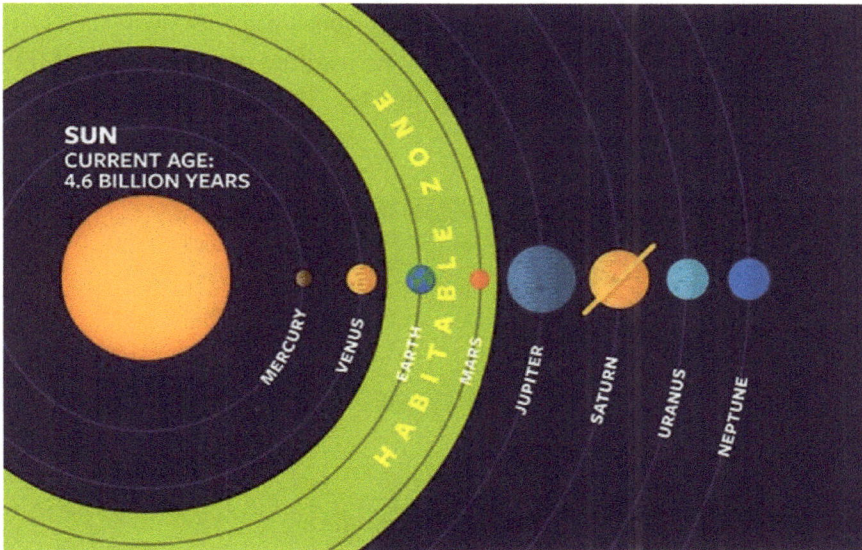

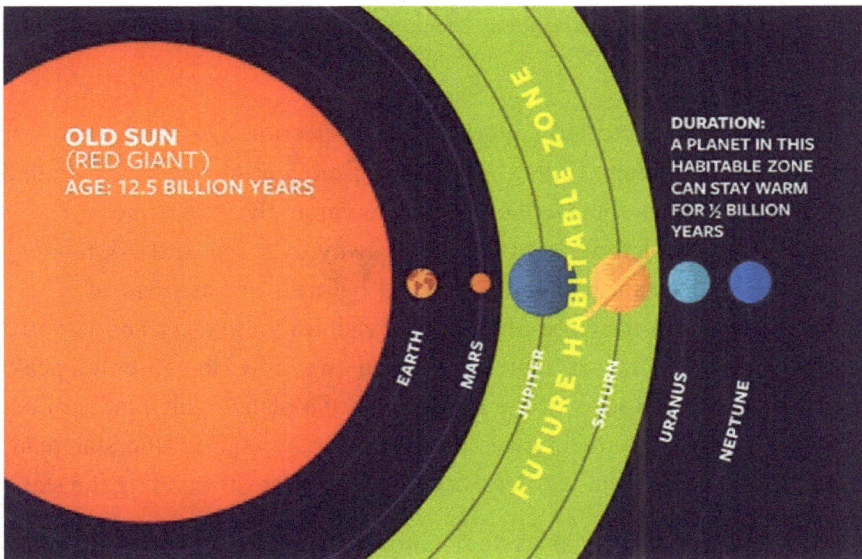

Fig. 1.18 Earth and Mars are currently enjoying their position in the Sun's habitable zone. However, the Sun will become a red giant in a no so-near future resulting in the "death" of Mercury and Venus and pushing away the solar system's habitable zone. Credit: Wendy Kenigsburg.

Although a different phenomenon, a star consuming its own planets is what Kishalay De from the Massachusetts institute of Technology (MIT) and a team of researchers observed in May 2020 and reported in their study, as shown in Figure 1.19.[31, 32]

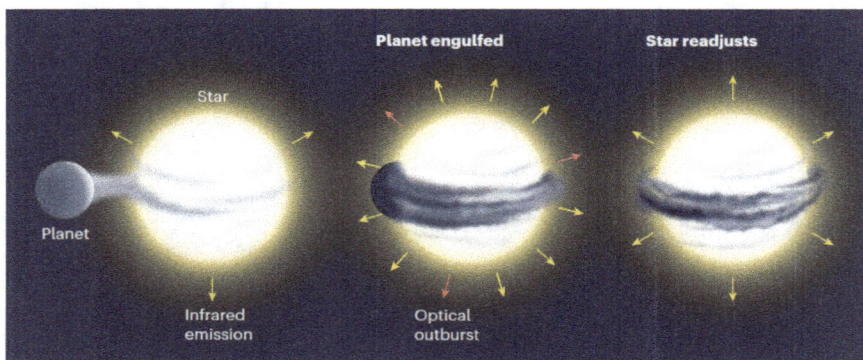

Fig. 1.19 A planet ten times the mass of Jupiter gets too close to its host star and gets swallowed causing a temporary increase in the brightness of the star.

These researchers detected a luminous mass ejection from a Sun-like star located approximately 12,000 light-years from Earth that lasted around 10 days. The outburst made the star become more than 100 times brighter over that period of time and then faded away over the next six months. That type of outburst has been observed before from mergers of binary stars. However, in this case, the optical (visible) luminosity and radiated energy, as well as the way luminosity faded away after the observed peak, are more of an indication of the engulfment of a planet with a mass of ten times the mass of Jupiter. After the planet was engulfed, the star read-justed itself, and all that was left was a trace of dust surrounding the star.

Coming back to the solar system, during the red giant phase, besides consuming some of its planets, the Sun's increase in size will also result in the relocation of the solar system's *habitable zone*, pushing it further away. Don't worry too much about this, as this won't happen any time soon. The Sun has been estimated to have consumed half of its hydrogen reserves during the first 5 billion years of existence. The Sun is in the middle of its main sequence phase and will need another 5 billion years to use up all its fuel. In other words, we are halfway there.

So, sleep well tonight, and let's continue with what happens to a star during its red giant phase. During this phase, the outer layers expand, but the core keeps contracting and increasing temperature. However, the star has not finished fusing hydrogen. This process now happens in the shell surrounding the core. We now have a helium core surrounded by a hydrogen-fusing shell. The core is now so dense that it becomes *degenerate.* Degenerate does not have anything to do with the moral or behavior of the star but is a way of describing that *electron degeneracy pressure* prevents the core from shrinking even more. This pressure results from electrons resistance to share the same space and is so strong that it counteracts the gravitational force.

When a star that is several times the Sun's mass has consumed all the helium in its core, it can reach high enough temperatures and pressures to trigger the fusion of heavier elements. This process can begin with the fusion of helium nuclei into carbon and then carbon nuclei to form heavier elements like oxygen, neon, and magnesium. However, not all stars are massive enough to achieve the conditions necessary for carbon fusion. These stars will end their lives after helium burning.

Planetary Nebula

For a star that has about eight times or less the mass of the Sun, fusion into heavier elements (larger atomic number) does not happen. The star begins to cool down and does not expand anymore. The star's outer layers are ejected, forming a beautiful planetary nebula like the one observed in Figure 1.20. At the center of the planetary nebula lies an object known as a *White Dwarf,* which is just a fancy name for the exposed core of the original star.

Fig. 1.20 The Cat's Eye planetary nebulae. Credit: NASA/ESA
Hubble Space Telescope.

Planetary nebulae have nothing to do with planets. They are just the ejected heated gas and dust remaining after the red giant phase. The name just stuck from the 18th century when astronomers noticed that these objects had a round, planet-like appearance.

White and Red Dwarfs

White dwarfs are made of extremely condensed material. A white dwarf usually contains about one and a half times the mass of the Sun condensed within a radius of around 10,000 kilometers or roughly the size of the Earth. If we could grab a teaspoon of white dwarf material, this would weigh about 15 tons!

As we mentioned previously, the Sun will take 10 billion years to consume all its hydrogen reserves.

For very-low mass stars such as *Red Dwarfs* (type M in the stellar classi-fication), those with masses less than half the mass of the Sun, the process by which they consume all their fuel can take way more than 10 billion years. Red dwarfs can take trillions of years to burn all their hydrogen fuel. We can look at these as reserved and cautious stars who play safe, low-risk players who spend their resources in a very conserva-tive way. They get to live long lives and will remain until the very end of the universe.

Once all the fuel has been consumed, white dwarfs and red dwarfs remain hot but continue to emit heat. Electron degeneracy pressure replaces gas pressure as the force operating against gravity, keeping white dwarfs stable for eons. It has been hypothesized that all the leftover heat will be radiated away to space after an extremely long time. As no more heat or light is emanating from these objects, dwarf stars will eventually become *black dwarfs*. Regardless of the inability to observe such objects due to the lack of emitted radiation, black dwarfs would still be able to be detected thanks to the effects produced by their gravitational fields. However, the time for a dwarf star to become a black dwarf has been estimated to be at least a hundred million billion years, which is way longer than the universe's current age. Hence, no black dwarf has been detected or will be detected any time soon.

Supernovae

The fate of massive stars, five times or more massive than the Sun, is completely different from that of lower-mass stars. These massive stars are the rebel ones. Those that live fast and die young. Like their low-mass cousins, these stars also fuse hydrogen into helium and subsequently fuse helium into carbon and oxygen. The difference is that they don't become merely red giants like their smaller cousins, but *red super giants*. The largest known red supergiant is thought to be VY Canis Majoris, with a size about 1,800 times the size of the Sun. The considerably larger atmospheres of massive stars are not the only implication of their initial larger mass. Gravity causes the cores of these stars to continue to collapse beyond what the core of a low and medium-mass star can collapse, reaching even higher densities and temperatures. This causes a succession of nuclear reaction sequences to take place, eventually transforming all the oxygen in the core into neon and eventually producing an iron core. Iron is the heaviest of the elements that can be produced through this mechanism. Iron fusion requires more energy than what it could give out after fusing; therefore, the star stops generating energy, causing the whole process to end. But now the fun begins. At this point, the temperature is so high that the nuclei of iron atoms are destroyed. This is a process called *photodisintegration*. Iron atoms are literally stripped down to individual protons and neutrons. The energy required to achieve this process is taken from the thermal energy of the remaining gas, making gravity the winner as there is not enough pressure to prevent further collapse of the core of the star.

Density continues to increase, and neutrons continue to get squeezed, to the point that the force until that point allowed those neutrons to stay together, the *strong nuclear force,* becomes repulsive, causing massive outward shock waves. These shock waves will eventually result in a magnificent explosion known as *Supernova.* Supernovae are among the brightest events in the universe. They are so bright, they can completely outshine the total luminosity of their galaxy. The most recent recollection of such an event occurred in 1604 AD. SN 1604, or the Kepler's supernova, named after the famous astronomer and mathematician Johannes Kepler, was so bright that it was even visible at daytime.

However, not all supernovae are created through the same process. Type Ia supernovae[33] is a particular type of supernovae that may occur on a binary star system comprised by a white dwarf and another stellar companion. At some point, the white dwarf starts stealing stellar material from its neighbor until it reaches a mass equivalent to 1.4 times the mass of the Sun. Once this happens, the white dwarf cannot sustain its own weight and explodes. Type Ia supernovae were instrumental in the discovery of the expanding acceleration nature of the universe in 1998 by two independent research teams[34, 35] which led to the award of the Nobel Prize in Physics in 2011.

Supernovae are not only useful in allowing humans to understand the universe's inner workings; these events are responsible for the production of elements heavier than iron. In the words of the late astronomer and science communicator Carl Sagan, "we are made of star stuff". A large portion of elements in our bodies, other animals' bodies, our environment, the solar system, were forged within the core and outer layers of a star during their lifetime. Depending on the stellar mass, atoms of elements such as carbon, oxygen, and iron, which can be present in the outer layers or the cores of a dying star, are ejected into the universe by supernovae explosions. With this material, new molecular clouds are formed, and through a process known as *neutron capture*, protons and neutrons within these newly formed clouds combine to form heavier elements, including gold.

The merging of neutron stars, which we'll discuss soon, is also responsible for creating approximately half of all heavy elements beyond iron. Neutron star mergers are also the only source of elements beyond bismuth (Bi) and lead (Pb).[36] By now, you must be agreeing with Sagan's statement.

One of my favorite versions of a periodic table was detailed by NASA,[37] explaining the origins of the elements. Only hydrogen (H), and helium (He) were created during the Big Bang, but the majority, 68%, of the known elements have been created majorly by star-related processes. Cosmic ray collisions, radioactive decay, and human-made elements are responsible for the rest. Note that while the Big Bang did create small amounts of lithium, NASA's table doesn't reflect this because most of the lithium we see today comes from dying low-mass stars.

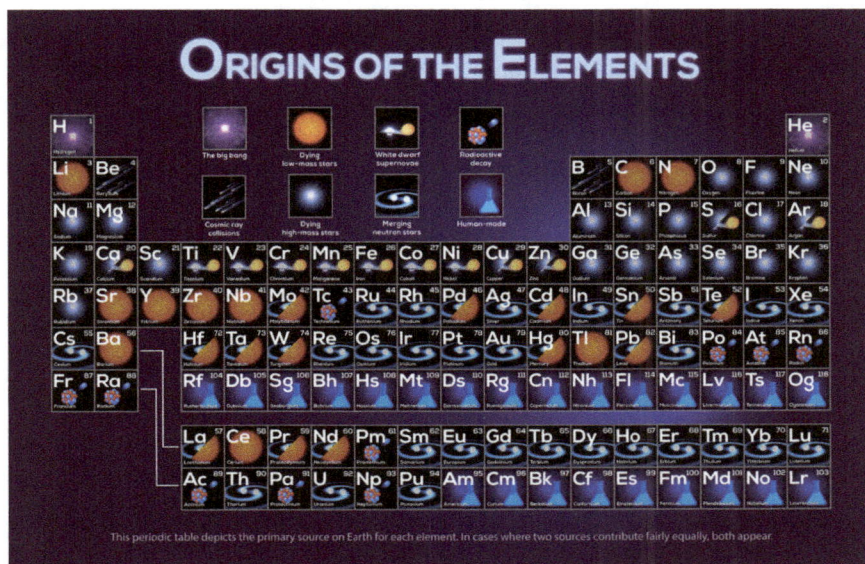

Fig. 1.21 The origins of the elements. Nearly 68% of all known elements are produced as a result of star-related processes. Credit: NASA's Goddard Space Flight Centre.

Neutron Stars

Once a star has gone supernova, what happens next? Well, it depends. For stars which initial mass was between 10 and 25 times the mass of the Sun, the remaining core will have a mass between one and a half and three times the Sun's mass and will continue to contract under its own gravity. Such a contraction causes electrons to fuse with the protons in the atom nuclei, forming neutrons, a process known as *electron capture*. The empty space within the atom shrinks at unimaginable scales. It is estimated that the radius of such cores will be roughly between 10 and 15 km, containing more than one and a half times the mass of the Sun. Matter here is compacted so much that the number of particles per cubic centimeter is extremely large. We now have a star which nucleus is entirely made of neutrons; a neutron star. A cube of a neutron star material of the size of a regular sugar cube would weigh more than 1 billion tons![38] Beyond this point, matter is not compressed anymore due to the *neutron degeneracy pressure,* which counterbalances the effects of gravity. Similar to the *electron degeneracy pressure*, neutron degeneracy pressure is the result of neutrons' resistance to occupying the same location. Such a resistance creates a force between the particles, which counters the inward pull of gravity, keeping the neutron star stable.

Now, let's go back to the principles we presented when we introduced our spinning dancer. As we discussed before, every time a spherical rotating object, in this case, the core of a star, reduces its radius due to gravitational collapse, its rotational speed increases. Therefore, neutron stars rotate incredibly rapidly. Their rotational speed is so high that the time they take to complete a full rotation is in the order of milliseconds. Let that sink in for a moment. This is an object of 10 km radius, roughly the size of Paris or Sydney, going through a complete circle around its center in milliseconds (one second has one thousand milliseconds). Earth completes a full circle around its center in 24 hours, which we call a day. Another interesting characteristic of neutron stars is their enormous magnetic fields. Even though the majority of particles within a neutron star are neutrons, there are still some remaining protons and electrons. Due to the high temperatures and pressures within a neutron star, these protons and electrons move at high speeds, generating electrical currents that result in strong magnetic fields. If Earth happens to be in the line of sight with these objects, that electromagnetic radiation could be detected. This is referred to as the "lighthouse effect", in the sense that, as the object rotates, the emitted beams of electromagnetic radiation will show up in the instruments as pulses of radiation from a specific source. Astronomers call such objects *pulsars*.

Pulsars

It is fair to say that all pulsars are neutron stars, but not all neutron stars are pulsars. Although all neutron stars have strong magnetic fields, astronomers might be unable to detect those periodic pulses from Earth if the stars are not properly aligned or the generated magnetic field is not strong enough or close enough to be detected. In those cases, these neutron stars cannot be classified as pulsars.

Fig. 1.22 Similar to a lighthouse observed from a distant ship (left), the electromagnetic radiation from a rotating pulsar (right) is detected on Earth. Credit: images generated by OpenAI's DALL·E.

The discovery of pulsars is quite interesting. In 1967, Jocelyn Bell Burnell, then a PhD student at the University of Cambridge, detected periodic pulses of radio emissions from specific sources in space at astonishing regularity. As this was something that no one had encountered before, and given the similarity with artificial signals in terms of their periodicity (like the electromagnetic signals humans use to communicate), Jocelyn and collaborators jokingly use the acronym "LGM" or "little green man" to designate their discovery. There is no indication that they were indeed thinking these signals were alien in nature, and it was just a practical joke they played…, or perhaps a strategy to attract the attention of the media and general public to their discovery.

Dr. Bell Burnell should have received the 1974 Nobel Prize for her remarkable efforts and discovery. However, her PhD advisor, Anthony Hewish, and Martin Ryle, who was the head of the Cambridge Radio Astronomy Group at the time, were awarded the prize. This occurred despite Jocelyn Bell running the telescope and analyzing the data that ultimately led to the discovery of pulsars. In an interview in 2021, Dr. Bell Burnell opined:[39] "the fact that I was a graduate student and a woman, together, demoted my standing in terms of receiving a Nobel prize." In my opinion, this is one of the greatest injustices in astronomy and science in general.

Okay, rant over.

The rotations of pulsars are so precise that they have been proposed as the main component of a potential interplanetary GPS system. The system *X-ray pulsar-based navigation and timing,* or XNAV technology was introduced in the 1980s, and the framework for the solution has matured enough in the last 40 years.[40, 41] Essentially, similar to how the GPS system works on Earth, the periodic X-ray signals from known pulsars are used to help determining the location of a spacecraft in outer space. When spacecraft are far from Earth, they don't have the luxury of a ground-based system to guide them. By constructing a database of known pulsars, which include their frequencies and locations, a spacecraft can compare all the received X-ray signals and calculate with a 2 kilometers accuracy its current location. Now, 2 kilometers accuracy might not be that impressive for you on Earth, but in space where distances are literally "astronomical", 2 km is a pretty good estimate. All this was theoretical until 1999, when the US Naval Research Laboratory launched a satellite experiment that demonstrated that a spacecraft can use pulsars to orient itself. In November 2016, China launched an experimental pulsar-navigation satellite, called XPNAV-1. XPNAV-1's objective was the Crab pulsar which is at 6,500 light-years from Earth in the constellation Taurus. The experiment's main goal was to verify the X-ray instrument's capabilities on board the Satellite. In addition, in June 2017, China also launched the Hard X-ray Modulation Telescope (Insight-HXMT) satellite, China's first X-ray astronomy satellite. Insight-HXMT science objectives include the observation of interesting X-rays sources such as black holes, neutron stars, and gamma-ray bursts. Insight-HXMT has also demonstrated X-ray pulsar navigation technology with great results.[42]

Fig. 1.23 An artistic impression of a spaceship employing an X-ray pulsar navigation system. Credit: Institute of High Energy Physics under the Chinese Academy of Sciences.

The most recent development in the X-ray pulsar navigation area took place with NASA's installation of the Neutron Star Interior Composition Explorer (NICER) device at the international space station in June 2017. NICER's main objective was to measure how big pulsars are to have a clearer understanding of the extremely dense matter they are made of. As an add-on, NASA also deployed the Station Explorer for X-ray Timing and Navigation Technology (SEXTANT).[43] SEXTANT timed X-rays signals from five different pulsars. Among those was the closest and brightest known millisecond pulsar, PSR J0437-4715, which is roughly 5,000 light-years from Earth in the constellation Dorado. This pulsar has a period of only 5.8 milliseconds; in other words, it rotates over its axis more than 170 times per second.

In NASA's words, SEXTANT could allow a spacecraft to "...triangulate their location, in a sort of celestial Global Positioning System (GPS), using clockwork-like signals from distant dead stars". The mission was capable of following each of the signals coming from the observed pulsars for about 5-15 minutes before autonomously positioning itself toward the next source. While orbiting Earth, the device measured tiny differences in the signal's arrival time and was able to calculate its own position in space without any human intervention.

Black holes

We have seen that neutron stars are extremely dense and are the result of stars with an initial mass between 10 and 25 times the mass of the Sun. What is the fate of stars with an initial mass even larger than 25 times the mass of the Sun? The answer is mind-blowing. Before getting into it, let's have a look first at a very famous physics term: the *escape velocity*. This is the velocity at which an object needs to move if it wants to escape (hence its name) the gravitational pull from a planet, moon, or any other celestial body without being pulled down to the surface. In the case of the Earth, an object needs to reach a velocity of 11.2 km per second, or 40,320 kilometers per hour. This means that an object requires a speed of minimum 11.2 kilometers per second to escape Earth's gravitational pull.

The escape velocity for a given celestial object depends on the celestial object's mass and the distance from the center of mass. For simplicity, let's consider a planet. The larger the mass of a planet, the larger the required escape velocity is. However, the escape velocity also depends inversely on the distance the escaping object is from the planet's center of mass. If we assume that the escaping object is in the planet's surface, the smaller the planet's size, the closer to the center of mass of the planet the object will be and the larger the escape velocity is when compared to a larger planet with the same mass. Surprisingly, the escape velocity isn't influenced by the properties of the escaping object itself. One might expect that escaping for a heavier object would be harder, but this isn't the case. This discrepancy is a classic example of how intuitive reasoning often doesn't align with physics concepts. The concept of escape velocity assumes that the escaping object has mass as it involves the conversion of gravitational potential energy—the energy an object possesses due to its position in the gravitational field of a massive object— into kinetic energy—the energy that a body possesses due to its motion, following the principle of energy conservation.

In this scenario, the farther away the object gets from the massive object, the higher its altitude, the more the object loses potential energy, and gains kinetic energy.

We have discussed how a star, during its final stages, gets contracted by its own gravity, closing more and more the empty space between the different components of the atoms that make up its core. In essence, the mass of the core of a star remains constant, but its size is reduced drastically. Keep in mind that a typical neutron star can have a radius of the size of a city like Sydney, around 10 km, but contains the matter equivalent to one and half times the mass of the Sun. What this means is that the more a star's core contracts, the larger the velocity required to escape the star's gravitational field.

In the late eighteenth century, the physicist Pierre-Simon de Laplace and John Michell, an English natural philosopher, independently came to the same question: can a star have a mass so large and a radius so small that the escape velocity would exceed the speed of light? But why focus on the speed of light? Well, most of us are aware—likely thanks to sci-fi movies—that the speed of light is the ultimate speed in nature, an astounding 300,000 kilometers per second or 1.08 billion kilometers per hour (with one billion being equivalent to one thousand million).

Photons are the fundamental particles of what light is made of, have no mass, and of course, travel at the speed of light. Scientists often refer to photons as *massless* particles. Photons are either produced by a source of light, such as stars, or reflected by a planet, a car, a dog, other people, etc. Photons hit the photoreceptors in the back of our eyes causing the photoreceptors to generate electrical signals. These signals are sent to our brains and then processed, resulting in the images we see.

Laplace and Michael hypothesized that if a star's escape velocity exceeds the speed of light, no photons would be allowed to escape, preventing them from reaching our eyes or instruments. Therefore, such stars should be completely dark. Michell called these then theoretical objects: *dark stars*.

With that brief history lesson, let's go back to the point in time after a star with an initial mass of more than 25 times the mass of the Sun has gone supernova. Similar to the fate of a neutron star, a very dense core will survive. The surviving core of these massive stars will have a total mass above three times the mass of the Sun. For neutron stars—which surviving core is about one and a half times the mass of the Sun—the neutron degeneracy pressure can stabilize the star's core, preventing it from contracting even more. However, at these larger masses, neutron degeneracy pressure is not enough to counteract the effects of gravity. The core then continues to contract, squeezing the material so hard that all the mass of the core will be contained on a single point or *singularity*, giving origin to a *black hole*. Similar to the error you get in your calculator when dividing between zero, the term singularity simply exemplifies that we don't know what happens at the interior of a black hole. The closer we would be able to be or even observe this singularity is the *event horizon* of the black hole. In colloquial terms, the event horizon is described as the edges of the black hole. Anything that passes through the event horizon even light itself, will be trapped forever, unable to escape.

But wait a second; we discussed how the escape velocity concept is based on the principle of conservation of energy in terms of the gravitational potential energy of the object trying to escape being transformed into kinetic energy. For an object to have potential energy, it requires mass, but did we not mention that photons are massless particles? How, then, can light (photons) be unable to escape the gravitational pull of a black hole? In the classical Newtonian framework, gravity is a force that affects objects with mass, but this does not make any sense. Photons should not be affected by gravity due to their massless nature.

According to Isaac Newton (1643-1727), gravity is a *universal force* which to whose intensity decreases with the square value of the distance from the source of the gravity. The "universal" term here refers to a force that reaches across the universe. Gravity from a source will be felt everywhere, but its intensity depends on how close or how far away an object with a given mass is from the source. However, gravity was redefined by Albert Einstein's theory of General Relativity in the early 20th century. In general relativity, gravity is described as the curvature of the spacetime caused by the presence of mass or energy. Therefore, the more massive an object is the more spacetime is warped around it. A typical way of visualizing this is to imagine an elastic membrane stretched across an elevated wooden frame.[44] This type of apparatus is represented in the following figure, and it is known as *gravity well*.

The elastic membrane represents the spacetime. If we place different objects on it, we can observe the deformation of the membrane. So, if we start with a ball like the ones used in a pool table game, we will be able to see how the membrane is distorted. If we then remove the first ball and use a bowling ball (greater mass), we will see a larger distortion. Even more, if we put both balls together, we will see that the smaller ball will be "attracted" towards the larger ball.

Fig. 1.24 A gravity well is commonly used to explain how an object with a given mass curves the spacetime continuum. Credit: Arbor Scientific.

As black holes are extremely massive, they curve the spacetime in such a way that light is trapped in that curvature, preventing it from escaping.

That's why, regardless of photons being massless, they are still subject to the enormous gravitational field of a black hole.

Black holes' physics is an exciting field of research. These objects were purely theoretical until the early 70s. This was until Thomas Bolton, a University of Toronto researcher declared that he had found an "invisible mass that was eating away at a giant blue star" in the constellation of Cygnus.[45] The object, Cygnus X-1, is an X-ray source located around 11,000 light-years away from Earth.

Since then, scientists have had confirmation of black holes through multiple ways, with *gravitational waves* and *direct imaging* the most exciting, at least for the general public.

Gravitational Waves

In 2016, scientists at the Laser Interferometer Gravitational-Wave Observatory (LIGO) announced the detection of gravitational waves caused by the merger of two black holes. Gravitational waves were predicted by Albert Einstein in his general theory of relativity more than a century ago. LIGO facilities are located in Livingston, Louisiana, and consist of two 4 km long arms. LIGO was built under the premise that extreme violent and energetic astronomical events produce gravitational waves, which, in principle, could be detected. Examples of such a violent and energetic events are the merger of two black holes or the merger of one black hole and a neutron star. These waves propagate throughout the fabric of space itself at the speed of light.

In LIGO, a laser light is sent into the instrument and a *beam splitter* splits the light into two identical beams sent along the 4 kilometers arms. The laser beams then hit a mirror that is located at the end of each arm and bounce back. When the light beams are back from their 4 kilometers long journey, the results are analyzed.

LIGO takes advantage of the very same nature of light. As we previously indicated, light is made up of particles called photons. However, as discussed, light also propagates as a wave; physicists refer to this as the *wave-particle* duality nature of light. A wave has peaks and valleys, and when two or more waves interact, they generate interference patterns.

Consider two waves, it is said that *constructive interference* has occurred when the peaks of the two waves (and the two valleys) coincide effectively creating a larger wave. On the other hand, if one wave's peak coincides with the other's valley, they cancel each other out; this is called *destructive interference*.

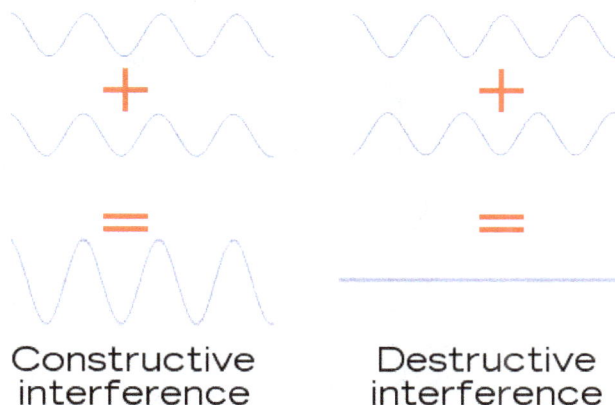

Constructive interference   Destructive interference

Fig. 1.25 Waves create different interference patterns when they interact with each other. Credit: Image by the author.

Instruments that combine two or more sources of light to create an *interference* pattern are called *interferometers*. Simply put, LIGO is a gigantic interferometer.

In LIGO, if there are no gravitational waves present, the detectors, which are located at one of the ends of each arm, will observe the bouncing beam of light at the same time, indicating that each beam traveled the same distance. This results in the bounced laser waves canceling each other out, generating a destructive interference pattern, and no resulting signal will be produced. However, gravitational waves cause space itself to stretch in one direction and simultaneously compress in a perpendicular direction. This translates into one of the LIGO arms becoming larger and the other one shorter in the presence of gravitational waves. Light now requires more or less time to bounce back after hitting the mirror at each arm's end. Depending on whether it is traveling through the now longer or shorter arm, a resulting signal or a *constructive interference* pattern will be observed.

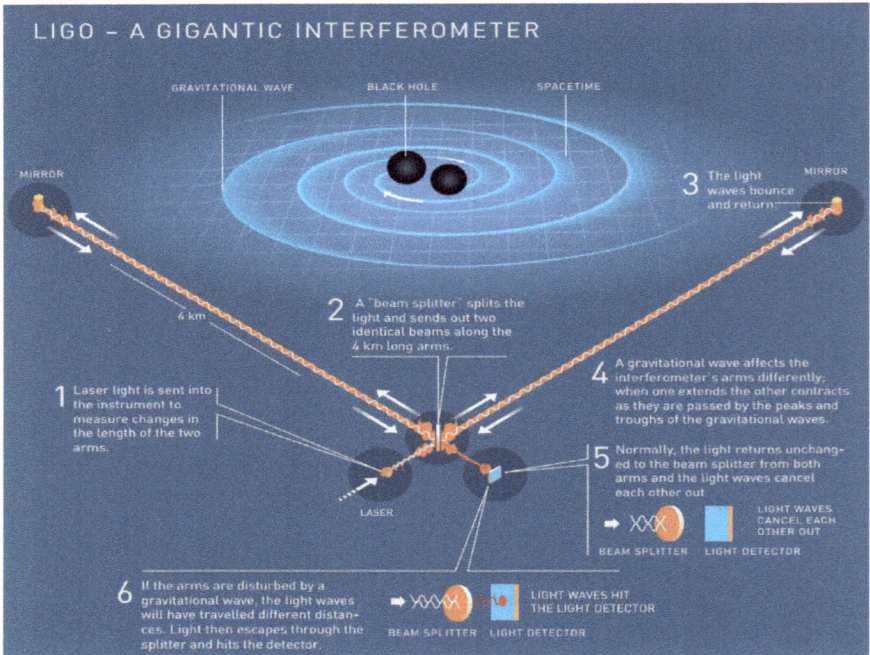

Fig. 1.26 A giant interferometer. LIGO has allowed the discovery of the elusive gravitational waves. Credit: LIGO.

Easier said than done. Measuring such tiny variations in the resulting signal is a monumental effort. The signal measured on September 14, 2015 was about one-thousandth the size of a proton! It has been reported that even Einstein himself was skeptical about humans being able to measure gravitational waves. Such an amazing accomplishment was recognized with the Nobel Prize in Physics award in 2017 to Rainer Weiss, Barry C. Barish, and Kip S. Thorne. As usual, these three remarkable people were only the visible faces of a large and extremely talented team of scientists and engineers all around the world.

The fact that astronomers now have the capability to detect events using gravitational waves opens a whole new chapter in astronomy and marks the beginning of the so-called *multi-messenger astronomy* era. In this new era, astronomers don't rely on a single *messenger,* such as light (visual, x-ray, etc.), to extract information from the universe.

Science has proved then that black holes are not just theoretical mathematical models but very real objects in space-time. What is the closest of these objects to Earth? Are black holes a threat to our existence? In universe scales, black holes are closer to Earth than what we can imagine. Thankfully, we don't need to lose any sleep about them.

Since the 60s, astronomers have known that most galaxies, including our home galaxy, the Milky Way, have supermassive black holes at their cores. Researchers categorize black holes as supermassive when they are in the order of hundreds of thousands or even millions to billions of times the mass of the Sun. The Sagittarius A* black hole is located at "only" 25,000 to 27,000 light-years from Earth right at the center of our galaxy. This is, therefore, one of the closest black holes to Earth. However, the closest black hole is located in the Gaia BH1 system, a binary system comprised by a G-type main-sequence star and a black hole. The black hole has a mass of 10 times the mass of the Sun and is located at a distance of 1,500 light-years from Earth.[46]

Black holes are traditionally detected via their effects on their vicinities. Particles spiraling inwards towards black holes form accretions disks that surround them. These spiraling inwards particles are subject to extraordinary gravitational forces, causing them to collide, rub, and bounce against each other. This results in frictional heating manifested as energy radiated away, which can be detected by the astronomers' instruments. The temperature of accretion disks of black holes is so high that some black holes can act as bright sources of x-rays, gamma rays (extremely energetic waves with the smallest wavelengths in nature), and visible light.

Direct Imaging (sort of)

All this is pretty cool; however, a picture is worth a thousand words, and an image is what the Event Horizon Telescope (EHT) team achieved in 2019. The EHT is a global network of synchronized radio observatories in which signals are combined to effectively create a telescope of the size of the Earth. The radio telescopes are located in France, Spain, Greenland, Chile, the United States, Mexico, and the South Pole. In April 2019, the EHT team revealed the first-ever image of a black hole, specifically, the image of the supermassive black hole at the center of the galaxy M87, located at 2.5 million light-years from Earth.

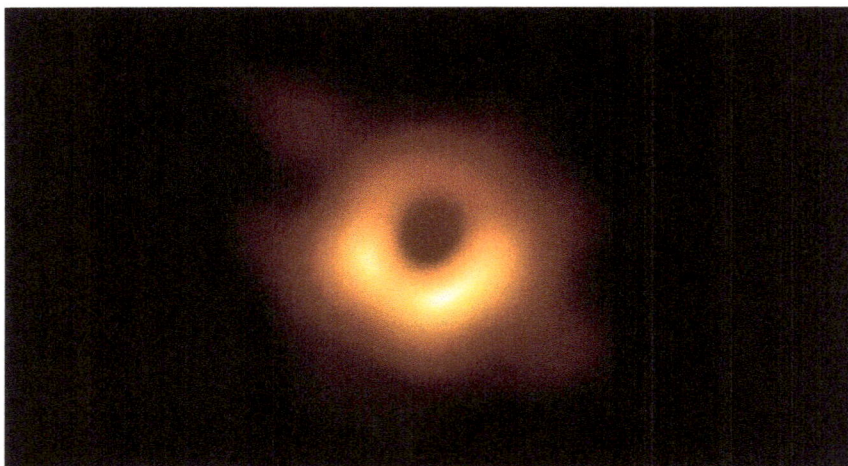

Fig. 1.27 Image of the event horizon of the supermassive black hole at the center of the M87 galaxy. Credit: EHT Collaboration.

The image is the result of decades of work by many talented people around the planet. How is it possible to effectively have a telescope of the size of the Earth? We mentioned before that signals can be combined to create a single signal using the technique called *interferometry*. Of course, this is not an easy feat and requires extreme precision. The EHT team is comprised of hundreds of scientists and engineers from 80 institutes worldwide. Atomic clocks were used at each telescope location to synchronize the different signals. These atomic clocks are so accurate that they won't deviate by a second in ten million years. The effort was massive from the engineering point of view, requiring the development of special digital signal processing techniques as well as data analysis and storage. The observation data was collected over five nights, from April 5-11, 2017. EHT Engineers have estimated that the total data volume collected was about 4 petabytes[47] (1 petabyte is equal to 1 followed by 15 zeros bytes).

To put this in perspective, Eric Shmidt, former Google's CEO, has estimated that the size of the Internet is roughly five thousand Petabytes. This means that the EHT collected an amount of data in five days equivalent to 0.1% of the whole data on the Internet. This may not sound too much, but remember that the Internet has been around for 40 years.

Recall that anything that has fallen into a black hole will be trapped forever. Therefore, to be clear, the picture taken is of the glowing matter in the accretion disk surrounding the black hole at the center of M87. Matter that is falling through the event horizon reaches speeds close to 30% of the speed of light[48] (astronomers call these types of speeds, relativistic speeds).

In May 2022, the EHT did it again. This time, the first image of the accretion disk surrounding the supermassive black hole at the heart of our galaxy, the Milky Way, was unveiled. Sagittarius A*, which is the name of our local black hole, has an estimated mass of "only" 4.2 million solar masses. This means that Sagittarius A*' mass is just a tiny portion of the previously imaged black hole at the center of M87, with a mass of 6.5 billion times the mass of the Sun.

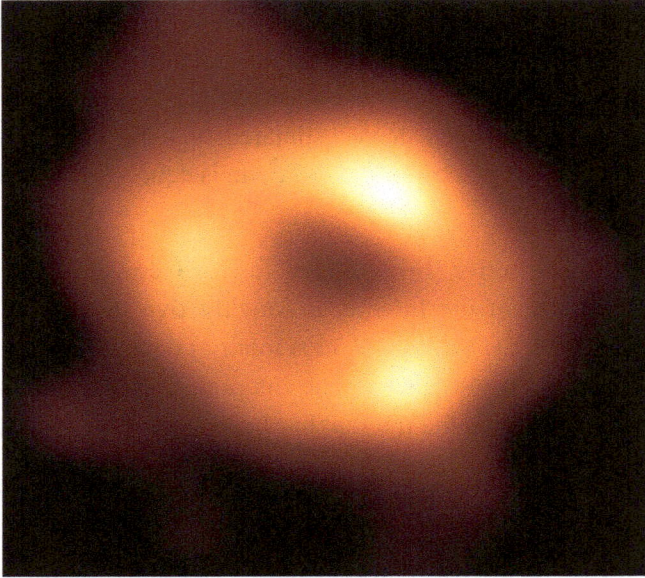

Fig. 1.28 Image of the supermassive black hole at the center of our own galaxy, the Milky Way. Credit: EHT collaboration.

As I was in the middle of writing this chapter, an astonishing nearly 33 billion times more massive than the Sun's ultra massive black hole was announced.[49] This monster, who inhabits the center of the galaxy Abell 1201, located about 2.7 billion light-years from us, was detected using the Hubble Space Telescope. The detection was possible due to a large bending of light caused by its enormous mass. Its mass is very close to the theoretical limit of 53 billion solar masses for a black hole.

Active Galactic Nuclei and Quasars

Why do supermassive black holes inhabit the center of galaxies? This is still a quite active area of research. However, some of the possibilities include that they have been formed as a result of the gravitational collapse of giant gas clouds from where galaxies originally formed shortly after the Big Bang. Other hypotheses indicate that they could result from the merger of many smaller black holes over millions or billions of years or the merging of supermassive black holes when galaxies collide.

Super and ultra-massive black holes give rise to what astronomers call Active Galactic Nuclei (AGN). AGNs are extremely energetic regions of some galaxies caused by the accretion of matter onto these black holes. The energy released by these events is so powerful that they can outshine the entire galaxy in which the AGN resides.

The most extremely luminous type of AGN is the Quasi-Stellar Radio Sources (Quasars). These are objects that are located at enormous distances from Earth and are characterized by the intense energy they radiate across the entire electromagnetic spectrum, from radio waves to X-rays. Quasars are amongst the most luminous objects in the universe, typically thousands of times brighter than the entire Milky Way.

Fig. 1.29 An artist's conception of a black hole emitting an X-ray beam. Credit: NASA/JPL-Caltech.

Regardless of supermassive black holes at the center of almost every large galaxy, the majority of black holes astronomers know of are the remnants of massive stars.[50] These objects are a testament that not even something as majestic as a star is immortal.

In this chapter, we have discussed how stars, much like living beings, are born, live their existence, and eventually die. From our experience, stars create the very conditions for planets to emerge and life to flourish. Life as we know it needs a planet, like the Earth. Whether progenies or siblings of stars, planets are essential to life. In the next chapter, we will explore the formation process of these objects and how stars determine their destinies.

Chapter 2
Planets

TL;DR

Our ancestors noticed night after night that certain lights in the sky did not behave like the rest. Five of such lights were the "uncommon" ones, which were moving against the background on a perceptible way. These lights were referred as "planētēs" or *wanderers* in ancient Greek; we call those nowadays, *planets*.

We have progressed from knowing the existence of only five planets to knowing that there are thousands. With the demotion of Pluto from planet to a dwarf planet in 2006, we only know of the existence of eight planets in the solar system. The other thousands of planets are located in different planetary systems and orbit around other stars; these objects are called *extrasolar planets*, *exosolar planets*, or simply *exoplanets*.

What is common to all planets is how they come into existence. They are remnants of the star formation process. By studying very young stars, astronomers have observed protoplanetary disks and gained insight into the real-time process of planet formation.

The location where a planet forms in relation to its host star, determines its composition. Planets that form closer to the star or in the inner side of the *frost line*, where temperatures drops to between 150 to 170 Kelvin (-123 to -103 degrees Celsius), are usually rocky. Mercury, Venus, Earth, and Mars are examples of these in the solar system. Planets that form in the outer side of the frost line and far away from the star's heat are usually giant gas planets. Jupiter, Saturn, Uranus, and Neptune belong to that category.

The *Core Accretion* model is currently the leading planet formation hypothesis. The model states that small objects accumulate gradually— accrete— due to gravity creating larger objects via collisions. Even after full-sized planets are formed, collisions continue to happen; even now. At the beginning of the solar system, collisions were more frequent and were responsible for a number of features we observe today, including the origin of Earth's moon.

Nowadays, pieces of other worlds, coming from regions populated by objects that are the remaining of failed or broken planets, continue to visit us, evidencing the ever-changing and active nature of the skies.

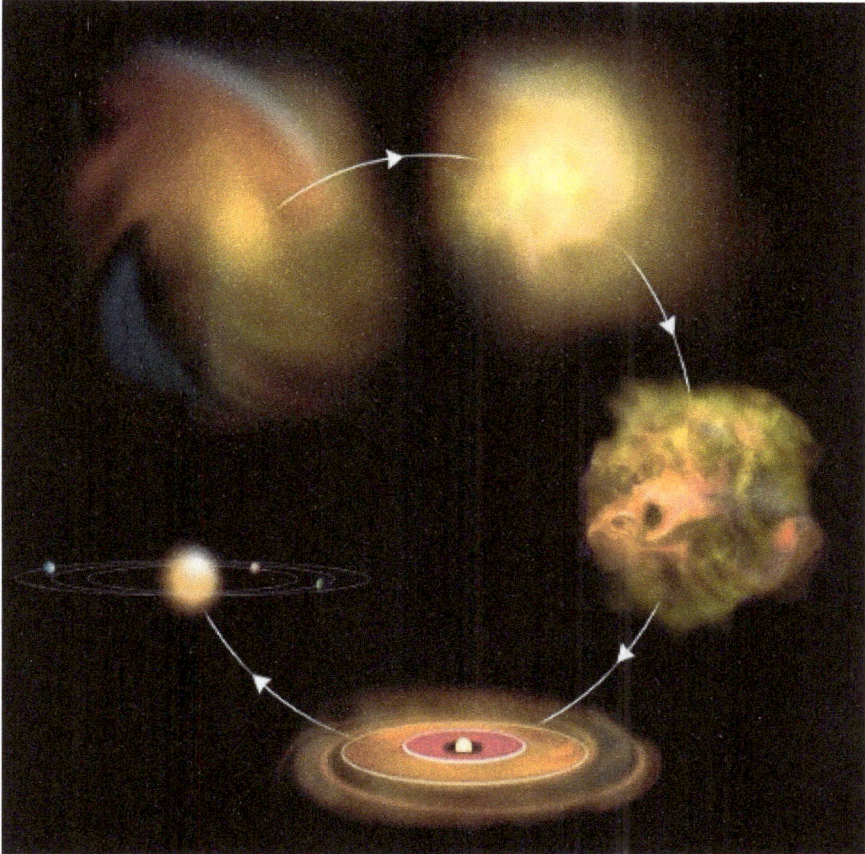

Fig. 2.1 Stages of the formation of a planetary system. A molecular cloud contracts, and flattens as it spins up. A star forms in the center and the leftover material may form planets. Credit: Bill Saxton/NRAO/AUI/NSF.

What is a Planet?

In our fast-paced world, we find comfort in knowing that some things remain constant over the years. One such constant is the fascination and awe inspired by the night sky; a sentiment many of us share. This has been the case since humans first roamed the Earth. Until recently, before city lights obscured the wonders of the night sky, people could observe a sea of stars and even bright gas clouds. Now, such a privilege is reserved only for those in remote and isolated regions.

While contemplating the night sky, our ancestors noticed that most of the lights they observed were static. For a long time people thought that the sky was invariant, immutable, and possibly, eternal. They did notice that out of all the static lights, there were five lights which, observed night after night, seemed to be "wandering" among the others. To differentiate those five lights from the rest, they called them "wanderer" lights. In ancient Greek, "wanderer" is written as "planētēs", a word that has survived the passing of time. The "Planētēs" word has had very little variation over the years in many languages: in English, "planet", in Spanish, "planeta", German, "planet" and so on. The five wanderer lights are now known as *classical planets,* and these are: Venus, Jupiter, Mars, Mercury, and Saturn. These are the planets that are bright enough, due to their proximity or size, to be seen without any optical instrument.

For ancient Greeks, these five wanderers were deities who they worshipped. Divine beings that were able to walk the heavens. However, worshipers of planets are not that common anymore (I personally don't know anyone, but maybe you do).

The "static" lights are stars and galaxies, which, as a matter of fact, are also moving. The distances at which these objects are from us are so large, however, that we cannot notice a change in their position night after night with our naked eye. On the contrary, planets in the solar system are "close" enough to Earth, and we can perceive their movement in smaller timeframes.

Eventually, the count of the number of planets in the solar system went from five to nine (including the Earth) with the discovery of Neptune in 1610, Uranus in 1781, and Pluto in 1930.

Depending on how old you are, you probably grew up listening to your teacher telling you that:

"A planet is a celestial body that orbits the Sun, and there are nine of them."

If you were born after 2006, your teacher would have said something along the lines of:

"A planet is a celestial body that orbits a star, and we have discovered hundreds of them. There are eight planets in the solar system...."

From the above sentences, you would have noticed two things: first, we used to say that there were only nine planets; that is not the case anymore. Those extra hundreds and now thousands or so planets are what astronomers call, *extrasolar planets* or *exoplanets* for short. An exoplanet is a planet that orbits a star that is not the Sun. We will talk more about how astronomers are able to detect those planets and some of the weird ones that have been found so far in the following chapters.

Second, what happened in the solar system? Did someone blow up a planet, so we ended up with only eight? Of course not. As far as I am aware, no one has been able to come up with some sort of a Death Star* weapon as depicted in the Star Wars movie franchise; someone, somewhere, must be trying for sure.

---

* A Moon-sized space station capable of destroying entire planets.

Fig. 2.2 The Death Star. Empire's ultimate weapon capable of destroying entire planets. Credit: Star Wars.

What happened is that the definition of a planet has changed. Humans tend to highlight the characteristics of things to classify them. *Categorization* is a way for us to simplify what we encounter in our day-to-day world and, more generally, our universe. By creating categories, we can recognize patterns, hence, helping us to understand the nature and origin of objects, people, animals, etc. Unless you have been living under a rock for the last 18 years, I am sure you have heard about the demotion of Pluto. The former planet, discovered in 1930, was "downgraded" to the category of *dwarf planet* on August 24, 2006, by the International Astronomical Union (IAU). There is even a Pluto demoted day, "celebrated" on August 24th every year.

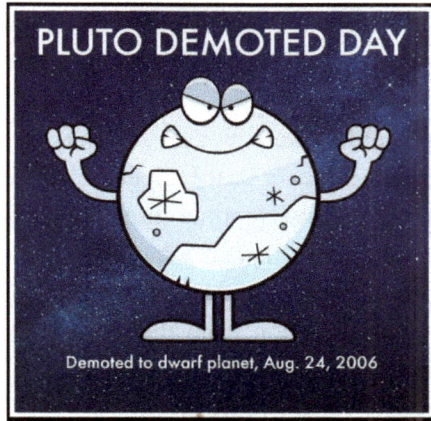

Fig. 2.3 A depiction of an angry Pluto after realizing that is not a planet anymore. Credit: the public Internet.

According to the IAU, a celestial body is considered a full-sized planet when it fulfils the following three criteria:[1]

1. The object orbits the Sun (or the host star). All objects orbit the Sun, but some of them don't do it directly, as is the case with moons, which orbit their parent planets.
2. The object has enough mass to sustain a stable shape and size over long periods of time. In other words, an object to be considered a planet must be round. A round shape indicates that the pull of gravity is the same in all directions.
3. The object has been able to "clear the neighborhood" on its orbit around the Sun.

Number three is what caused Pluto to be "demoted". Pluto has not been able to remove other bodies of comparable size (excluding its own satellites) from its orbit around the Sun. Pluto is located in what is called the *Kuiper Belt*. A disc-shaped region that extends about 30 to 50 astronomical units from the Sun. The Kuiper Belt region is named after Dutch-American astronomer Gerard Kuiper (1905-1973), who in 1951 predicted its existence. The region is populated by many small, icy objects of comparable size to Pluto and other smaller bodies known as Kuiper Belt Objects (KPOs). Pluto shares its orbit with many of those Kuiper Belt objects, therefore, failing to fulfill criterion number three of the International Astronomical Union. The main reason for demoting Pluto is that there are other five dwarf planets in the Kuiper Belt, and the existence of many more is quite likely. For instance, Triton, Neptune's largest moon, is believed to have been a past Kuiper Belt object. Triton is larger than Pluto and currently orbits Neptune in a retrograde fashion (rotating in the opposite direction from that in which the planet rotates about its axis). This, added to a highly inclined orbit of approximately 23 degrees, suggests Neptune captured Triton from the inner Kuiper Belt.[2] Triton is important in the search for life in the solar system, as we will explore in Chapter 7.

I believe that, given the possibility of the existence of tenths of objects of a similar size or larger than Pluto in the Kuiper Belt, it is not practical to consider them planets. Such an approach would potentially end us up with tenths of planets in the solar system. Imagine those poor kids having to memorize all those names at school.

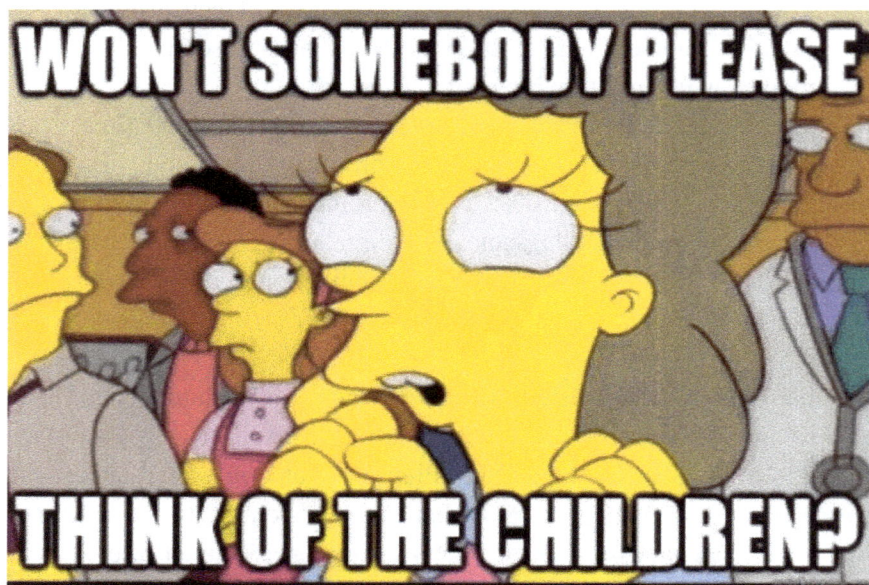

Fig. 2.4 Kids at school would need to memorize thousands of planet names. Let's think of them. Credit: meme from the public Internet.

Does Pluto care about this? Not at all. Pluto will continue to exist, rotate, orbit, etc., as it has been doing for billions of years. Classification of things is just a human construct. Potential extraterrestrial astronomers may have a different definition of what a planet is.

The New Horizons mission[3] arrived at Pluto in 2015 and took some amazing pictures. New Horizons is the first mission to fly to the Pluto system (Pluto and its companion dwarf planet Charon), and the Kuiper Belt.

New Horizons provided a detailed snapshot of the current state of Pluto's and Charon's atmospheres and details of their surfaces. The diverse features observed, which include mountains made of water and ice and evidence of tectonic activity on Charon, among other discoveries, demonstrate geological diversity and activity in these distant worlds.[4] Such observations only helped to confirm the status of Pluto as a dwarf planet.

Fig. 2.5 Pluto here completely unaware that humans don't consider it a planet anymore. Credit: NASA/New Horizons mission.

But don't feel too disheartened by the news about not having nine planets in the solar system anymore. Astronomers have been observing anomalies in the orbit of objects that reside in the outer solar system, specifically, objects that are located beyond the orbit of Neptune. These trans-Neptunian objects (TNOs), as they are known, exhibit a series of anomalous patterns that cannot be explained by Neptune's gravitational influence or attributed to residual dynamics of the formation of the solar system. These anomalies, however, seem to hint at the existence of a yet-to-be-discovered massive planet known as *Planet Nine* (P9), whose gravitational influence could be playing an active role in the landscape of the trans-Neptunian space.[5] Recent simulations and statistical analysis indicate the existence of Planet Nine with a high confidence.[6] These simulations also suggest that this planet would have a highly elongated orbit, a mass of about five Earth masses, and an average distance of 500 AU from the Sun. Its highly elongated orbit means that it would take this planet between 10,000 to 20,000 years to complete one full orbit around the Sun. After all, we may end up with nine planets in the solar system.

Fig. 2.6 An artist's conception of the hypothetical Planet Nine located beyond the orbit of Neptune. Credit:NASA/Caltech/R. Hurt (IPAC).

According to the International Astronomical Union (IAU), we have learned what's considered a planet. I have to admit that their definition is pretty academic, or in other words: boring. However, one of my favorite definitions reads: planets are just the leftovers of the star formation process.

Protoplanetary disks

In Chapter 1, we discussed how our current understanding indicates that stars are formed when a molecular cloud, also known as a nebula, contracts. We described how gravitational instabilities cause gas and dust grains to collapse gravitationally. Nebulae are not uniformly dense and, on the contrary, many regions are denser than others. In each of those regions, material is gravitationally pushed to the center, eventually forming a star. However, material that has not been used to construct the star, the leftovers, will also group in regions known as *coagulation zones*.

Similar to the concept of *protostellar disk*, which refers to the disk-like structure present before the star is fully formed, we can also talk about a *protoplanetary disk*. A protoplanetary disk is a disk surrounding the protostar where residual gas and dust is found. Unfortunately, these disks are pretty dark in the sense that they don't reflect too much of the light from the protostar, and therefore, they are difficult to observe in the visible light portion of the spectrum. In addition, dust particles and gas receive the light from their host young star and re-emit it in millimeter and submillimeter wavelengths. Due to the properties of our atmosphere, light with these characteristics can only be observed on Earth, at high-altitude and dry locations. Water vapor in the atmosphere prevents these wavelengths from reaching the ground. The lower the altitude of a given location, the greater the density of the atmosphere is, and therefore, the greater the concentration of water present. This makes it extremely diffi-cult to detect and observe these millimeter and submillimeter wave-lengths of light at places of lower altitudes.

At high-altitude dry-environments locations, the density of the atmosphere is lower, and there is not too much water vapor in the air. Such conditions make less possible for Millimeter/submillimeter signals to be absorbed by the Earth's atmosphere, allowing them to reach our observation instruments. This is why the Atacama Large Millimeter/submillimeter Array (ALMA) observatory was built in the Atacama Desert and why this instrument is the ideal ground-based tool to image protoplanetary disks. The Atacama Desert sits at 2,500 meters above sea level and is the driest non-polar desert in the world. Some areas receive less than 1 millimeter of rain per year. The ALMA observatory is an interferometer comprised of 12-meter antennas separated by 16 kilometers from each other and observes at wavelengths between 0.316 and 3.57 millimeters (mm).

Have astronomers observed protoplanetary disks? Yes, they have. In Figure 2.7,[7,8] we can see some of the images captured by the ALMA observatory as part of the Large Program Disk Substructures at High Angular Resolution Project (DSHARP) survey. Astronomers call a *survey* the systematic and exhaustive period of observing a significant portion of the sky. DSHARP is a survey in the 1.25 mm wavelength of 20 nearby, bright, and large protoplanetary disks. The aim of DSHARP is to characterize the different small-scale substructures present in these disk materials and determine their role in the planet formation process.

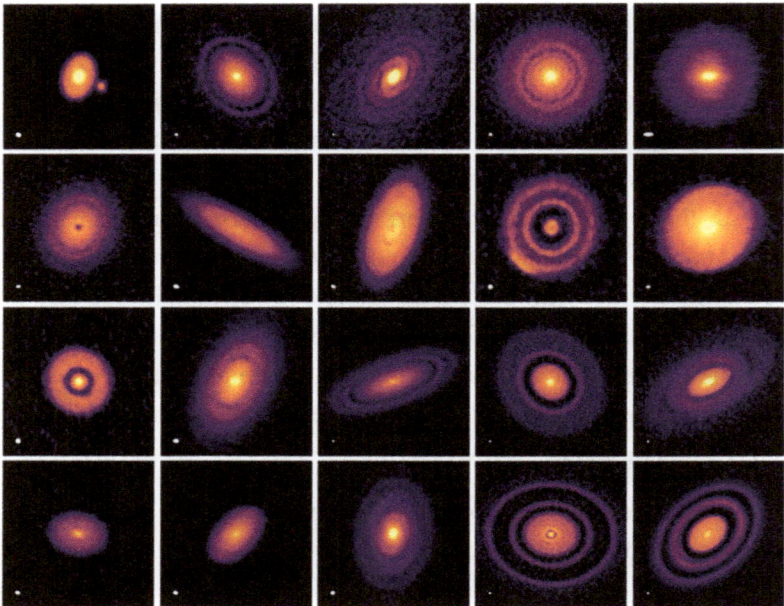

Fig. 2.7 Images from 20 nearby protoplanetary disks which are part of the DSHARP survey.

All these images show a bright point in the middle of the disks. The brightness of this point can indicate whether it's a protostar (a star where fusion hasn't started yet) or a young star, depending on the disk's evolution stage and the central object's age. On the other hand, the gaps that are observed might indicate the presence of one or more forming planets or *protoplanets*. As a forming planet accumulates —accretes— material, its gravitational field grows, interacting with the material in the disk. Although gaps in protoplanetary disks may be the result of other processes such as magnetic fields, turbulence, or gas being evaporated by the intense radiation of a close-by star —a process known as photoevaporation, they are mostly due to a *protoplanet* clearing up its orbit and establishing its presence in the nascent planetary system. Pay attention, Pluto; that's how it is done!

But how can this whole forming-a-planet process come to be? Can all those pieces of dust and gas get together and form a full-sized planet? First, we need to talk about what type of planets we have observed out there and some examples of such planets here in our own backyard, the solar system. Astronomers divide planets in the solar system into two large categories: rocky planets and gas giants. Rocky planets are closer to their host star, the Sun. Listed from their proximity to the Sun, these are: Mercury, Venus, Earth, and Mars. Further away, we have the gas giants, Jupiter, Saturn, Uranus, and Neptune.

Compared to the total mass of a planet, rocky planets have thin or non-existent atmospheres, while gas giants have extremely thick atmospheres, so thick that they constitute the majority of their total mass. For instance, Jupiter's atmosphere totals 96 to 97 percent of the planet's mass, which is equivalent to 305 to 308 times the mass of Earth. In comparison, Mars, which is a rocky planet with a mass of only 10 percent of Earth's mass, has a very thin or no atmosphere at all.

Planet formation models

So, how do planets form? Currently, there are two leading hypotheses: The Gravitational Instability model, and the Core Accretion model.

The Gravitational Instability model is considered a 'top-down' approach, or in other words, planets form big and then start contracting. The model indicates that dense clumps of gas form in the original proto-planetary disk. These dense clumps of gas or denser regions lead to self-gravitating structures (similar to the process of star formation in molecular clouds). Those self-gravitating structures continue to collapse under their own gravity. Their cores, however, are solid due to dust sedimentation within the gas; gas giants could have been formed in this way.

A helpful visual analogy to understand this (not fully comprehensive as the process is much more complex) is to imagine some sugar (solid particles) in a glass of water (a fluid). If you stir the sugar with a spoon, you can no longer distinguish the sugar particles in the water. After some time, some of the particles will dissolve in the water, but others will sediment at the bottom, similar to how the solid core of a gas giant sediments at the center of the planet.

The gravitational instability model has some drawbacks. For instance, it does not explain the formation of small rocky planets like Earth. In addition, in most protoplanetary disks that astronomers have had the chance to observe, massive planets are not present in the regions closer to the host star. Massive planets close to their host star are typically found in more mature systems.

Do astronomers have a model that explains what they observe better and how rocky planets form? Yes, they do. Enter *The Core Accretion model*. Contrary to the Gravitational Instability model, this approach assumes a 'bottom-up' approach. Microscopic solid dust grains and ice particles collide and stick together, forming larger particles of centimeters in size. Via the same mechanism, these centimeter-sized particles collide and stick together to form objects between 1 and 1000 kilometers in size; these objects are called *planetesimals*. Again, gravity makes these planetesimals stick together, forming *planetary embryos*. These planetary embryos eventually evolve into planetary cores or *protoplanets*. This whole process can take between one hundred thousand years and one million years. Formation of gas giants results when the conditions allow a planetary core's mass to grow enough —potentially between five and ten times the mass of the Earth— to accumulate the gas remaining in the protoplanetary disk. The more massive the core is, the more gas it can accumulate. This is due to an old friend, *the escape velocity*. Gas molecules have mass, and the higher the core's mass of a planet is, the larger the speed at which the gas molecules surrounding the core need to escape the planet's gravitational pull and escape to space. That is, the more massive the core, the more gas molecules the planet can retain, and therefore, the thicker the resulting atmosphere will be. Giant gas planets are usually formed on an average of between five and ten million years. Any remaining gas is dissipated, and dust particle leftovers are either dispersed by radiation from the star, destroyed in collisions, become part of other planetesimals, or even fall to the parent star due to the star's strong gravitational field.

The core accretion model explains quite nicely the origin of terrestrial and gas giant planets. Simply put, a planet with a more massive core will accumulate more gases, resulting in a thick atmosphere, whereas a less massive planet will not accumulate that much gas, causing it to end up with a thinner atmosphere or even not having an atmosphere at all. However, despite sharing a similar formation process according to the core accretion model, rocky and gas giant planets are not made of the same stuff. Astronomers use chemical elements to describe the composition of planets, stars, and any other object in the universe. This is by no means an easy feat, given the number of chemical elements in the universe. It is quite common, then, to classify planets based on the percentage of metals (any element other than hydrogen and helium, as we discussed in Chapter 1) present in their internal structures.

The Ice Line

As we have been discussing, rocky planets are close to their host star, while gas giant planets go further away; the closer you are to the central star, the hotter it is. At these high temperatures, water molecules cannot remain liquid or in any solid form. In addition, solar wind from the recently born star expels those molecules, along with helium and hydrogen molecules to the other regions of the planetary system. Therefore, rocky inner planets are made of metal and rock. At the outer regions, the climate is different. Farther away from the host star, the temperature drops considerably. This permits water to remain frozen, allowing giant planets to contain ice particles that behave like rocks. Giant planets then are made of metal, rocks, and ice. This extra ice component helps giant planets to be more massive than its terrestrial rocky counterparts, increasing their total mass and allowing them to hold onto the gas surrounding their cores. The distance at which this distinction in temperature from the host star happens is called *the frost line*, *snow line,* or *the ice line.*

But where is this *ice line*? Well, it depends on the host star. The hotter the star, the further out this line will be. It is usually located at a distance where the temperature drops between 150 to 170 Kelvin (-123 degrees Celsius to -103 degrees Celsius). In the solar system, this ice line is located about 4.5 to 5 astronomical units,[9] which is roughly the distance between the Sun and Jupiter. At this distance, compounds such as water, carbon dioxide, carbon monoxide, ammonia, and methane can condense into solid ice grains, which can accrete due to gravity, creating larger objects. Larger objects also gently merge if they are moving slowly or by hard impact if their speeds are not that slow.

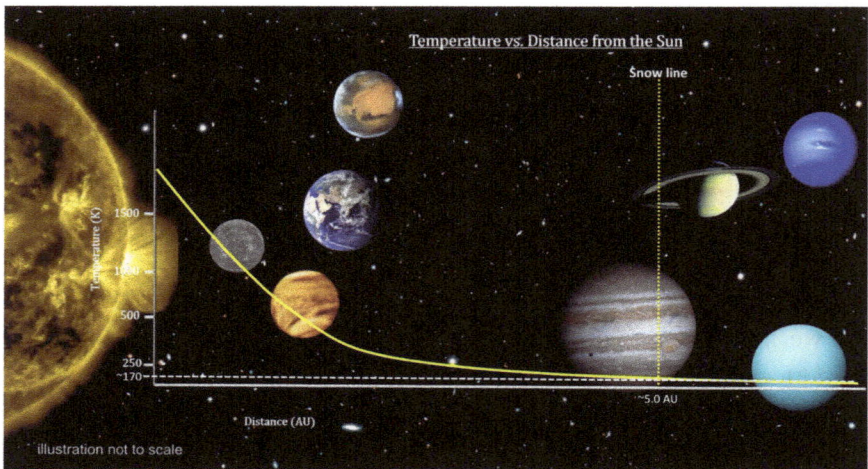

Fig. 2.8 In the solar system, the ice line is located about five astronomical units from the Sun. Credit: Image adapted from.

Visual evidence that supports the Core Accretion model was collected by the New Horizons mission[10] on the 1st of January 2019. I know! Science does not take holidays, and unfortunately, many scientists do not either. On that day, the spacecraft captured an image of the Kuiper Belt Object *2014 MU69*, or Arrokoth (Native American word for 'sky'). The object, a snowman-shaped space rock located in the Kuiper Belt, is the farthest object ever visited by a spacecraft. Arrokoth is a 36-kilometer-long body. Its age, based on the number of impact craters on its surface, has been estimated to be greater than 4 billion years.

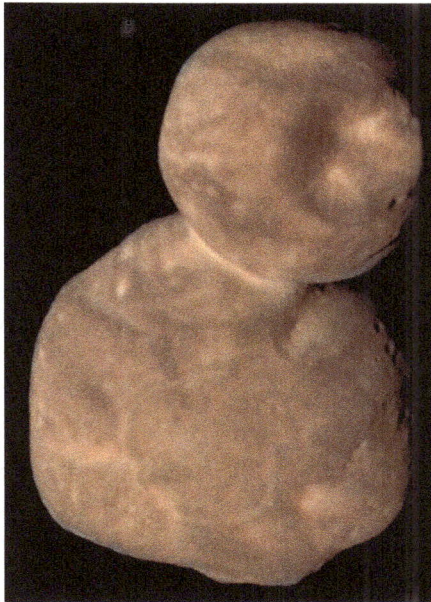

Fig. 2.9 Arrokoth (Native American for 'sky') resembles a snowman. The object is the result of accretion by gentle impact. Credit: NASA/ Johns Hopkins/ Southwest Research Institute/ Roman Tkachenko.

Arrokoth is comprised of two main lobes, each of which would have been formed by the aggregation of dust at the beginning of the solar system. The neck, or the joint between the two lobes, is well-defined and smooth, which supports the idea that the once-separated objects were rotating slowly before gravity bounded them together in a gentle collision. Despite this discovery not ratifying 100 percent the Core Accretion model, I am sure the model supporters were not completely unhappy with these findings. Arrokoth is extremely valuable as it is considered an object that has been present in the solar system since its beginning. Being so far away from other planets and the Sun itself has allowed it to be little disturbed. Objects like these are surviving relics that can help astronomers understand better the origin of the solar system.

Type-I and Type-II migration

It is set, then? Are astronomers one hundred percent sure that's how planets form everywhere in the universe? No way! One major problem with the core accretion model is that according to simulations, gas giant planets would have needed between 10 million and 1 billion years if formed via the gradual accretion of material. The main issue here is that it looks like planetary nebula lifespan is only around 10 million years. So essentially, there seems not to be enough time for gas giants to accumulate all their mass in such a short time. However, the presence of certain types of vortices in the protoplanetary disk (anticyclonic vortices) might explain how a planetesimal can increase the rate at which gas is accreted onto the core.[11] Of course, more observations and simulations are needed to validate this idea.

The second issue is the fact that this model predicts the formation of gas giant planets at large distances from the parent star. However, multiple observations have confirmed the presence of giant planets close to their host star. Astronomers even have a name for those planets: Hot Jupiters. We will talk in more detail about those in Chapter 5. So, how is this possibly explained? The current answer is: migration. Planets may indeed be formed where the core accretion model predicts, but during their lifetimes, they might migrate inwards, closer to the star, or outwards, even farther away from where they were formed originally. When planets are still forming, there is plenty of gas surrounding them. Smaller planets interact with that gas, causing them to lose angular momentum. As a result, the planet may spiral inwards towards the parent star. This is known as Type-I migration. If the interaction is enough to disperse the gas, the planet is saved from a horrible death and remains in a stable orbit.[12]

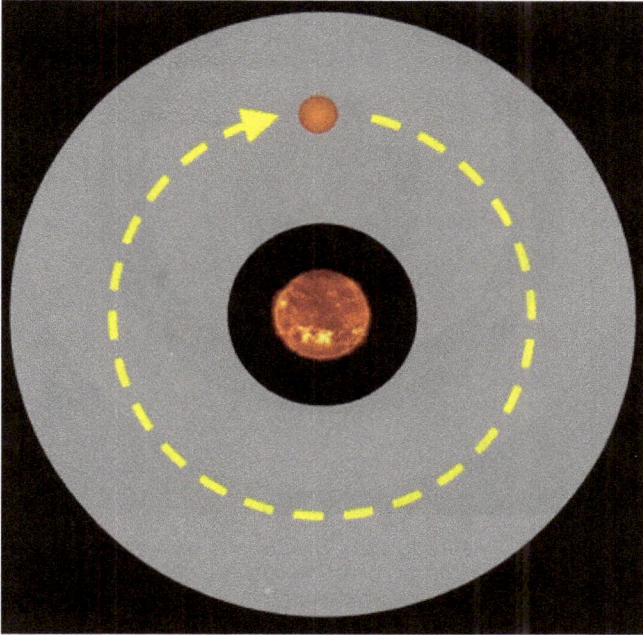

Fig. 2.10 Type-I migration. The planet is embedded in the gas disk and is unable to create a clear gap. Credit: @AstroPhil2000.

Larger planets experience something different. Larger planets around the size of Saturn can perturb even more the surrounding gas to the point that they create gaps in the protoplanetary disk. This means that the planet effectively splits the disk into an inner disk, closer to the star, and an outer disk, farther from the star. If the planet is closer to the inner disk, which is migrating inward towards the star itself, then the planet will share the same fate and migrate inward. If the planet is closer to the outer disk, which is getting away from the star, the planet will migrate outward away from the star, too. This type of migration is known as Type-II migration.[13]

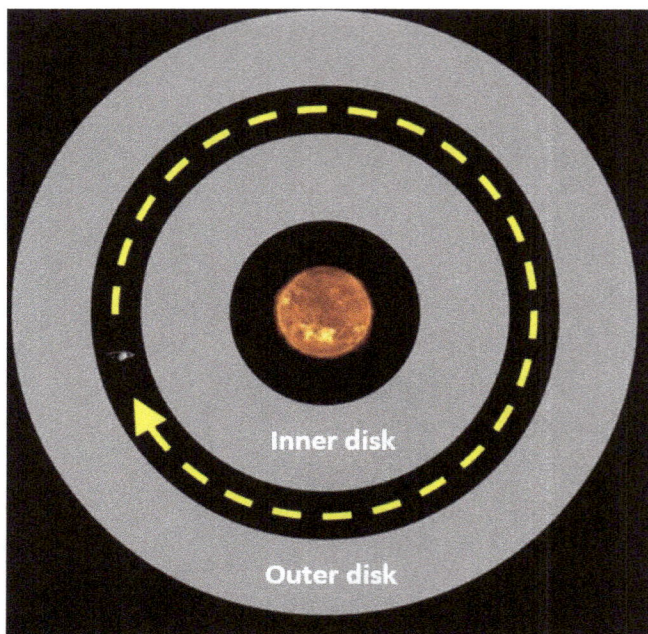

Fig. 2.11 Type-II migration. The planet is large enough to cause a gap in the gas disk. The direction of the migration depends on the location of the planet. Credit: @AstroPhil2000.

Type-II migration is quite important to the field of exoplanets. Hot Jupiters is the category to which the first exosolar planet, discovered in 1994, belongs. The discovery of a large planet orbiting very close to its host star was so surprising that its discoverers hesitated to report their findings. This is also evidence that planets must "halt migration" at some stage; otherwise, astronomers would not be observing these Hot Jupiters. Suggested planet migration halting mechanisms include, for instance, the planet entering a gap in the inner accretion disc. Such a gap could have been formed as a result of stellar wind or stellar magnetic field. On the other hand, strong tidal interactions between the planet and its host star could affect the distance from the star and the eccentricity—how much an orbit deviates from being circular— until the planet's orbit becomes stable.[14]

Such individual mechanisms, or a combination of those, can stop the migration of the planet at some distance.

But what happens after planets finish migrating? Do the orbits of the objects of a planetary system settle, and is the formation process over? Well, similar to when someone finishes a manual project, such as a model building, there are remnants, leftovers: pieces of unused cardboard here, sheets of paper there. You get the idea. If this analogy holds, astronomers should see evidence of such leftover materials in the solar system; well, they do, indeed they do.

Fig. 2.12 Leftover material can be seen after finishing the assembly of a model. In the same way, leftovers from the formation of the solar system can be observed nowadays. Credit: Image created by OpenAI's DALL·E

The Leftovers of the Formation of the Solar System

Several features in the solar system are a testament to planet formation and are evidence of its chaotic beginning. Two regions in particular, the *asteroid belt* and the *Kuiper Belt* populated by objects ranging in sizes from small rocks to dwarf planets remind us that the solar system was a very different place during its origins. In these regions, planet formation could not complete, possibly due to the gravitational disturbances of larger fully formed planets.

The Asteroid Belt

The *asteroid belt*, a region between the orbits of Mars and Jupiter, is located approximately between two and three astronomical units from the Sun, which means that it is placed on the inner side of the solar system's snow line. The objects in this region are rocky in nature, with Ceres being the largest known, with 940 kilometers in diameter.

Asteroid belts are not unique to our planetary system. In 1983, astronomers discovered a structure resembling an asteroid belt around the star Fomalhaut, a young star located about 25 light-years from Earth. However, 40 years needed to pass before astronomers could obtain an actual image. In March 2023, researchers using the Mid-Infrared Instrument (MIRI) in the most advanced telescope to date, the James Webb Space Telescope (JWST), were able to image[15] this asteroid belt revealing an extended inner disk, an inner gap, an intermediate belt, and a structure within a region that is free from the presence of asteroids. This free-asteroid region is known as the Kirkwood gap or KBA region, which is also present in the solar system.

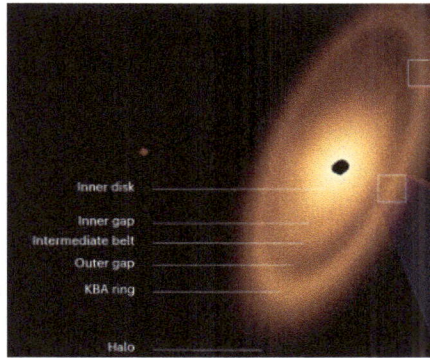

Fig. 2.13 Image of the first ever outside the solar system asteroid belt captured in 2023, forty years later after its discovery. Credit: NASA/ESA/CSA/A. Pagan/A. Gáspár.

The Kuiper Belt

The second region, the Kuiper Belt, is located well beyond the snow line, even past the orbit of Neptune, and is distant from the Sun. The objects in this region are primarily composed of rock and ice, with structures likely consisting of a rocky core enveloped by an icy mantle. The largest of these objects are the two dwarf planets, Pluto and Eris, with diameters of 2,377 and 2,326 kilometers, respectively.

Fig. 2.14 The asteroid belt and the Kuiper Belt. Leftovers of the inner rocky and outer gas giant planet formation. Credit: NASA.

Even farther, at distances from the Sun between 2,000 to 200,000 AU (0.32 to 3.2 light-years), researchers also find evidence of the solar system formation.

The Oort cloud

The cloud-like structure, known as the *Oort Cloud*, is a region where a group of weakly bound to the Sun icy objects reside. The cloud is named after the Dutch astronomer Jan Oort, who proposed the idea in 1950.

The Oort cloud concept aims to explain the origin of long-period comets.

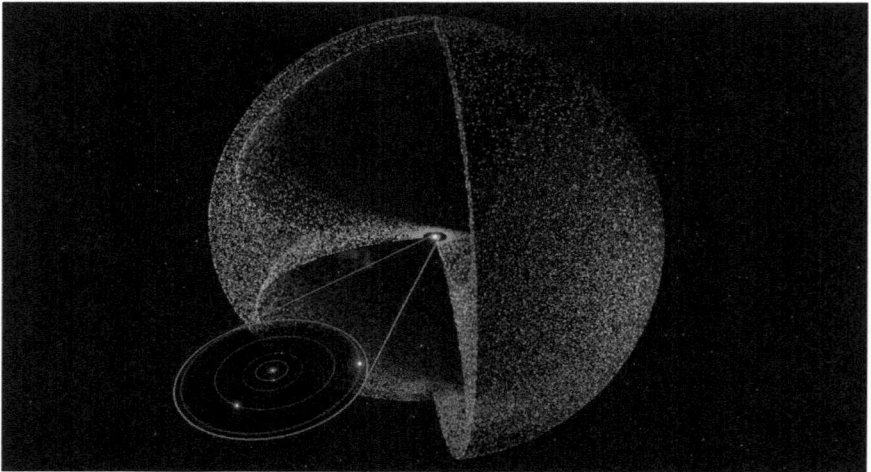

Fig. 2.15 The Oort cloud. A spherical layer of icy objects surrounding the Sun. Credit: NASA.

Comets

Comets are icy bodies formed from the original solar nebula in the outer regions. Many of such icy bodies were incorporated into the giant planets while some of them were ejected from the still-forming solar system or kept in reservoirs past the orbit of Neptune. Usually referred to as "dirty snowballs", they are mostly comprised of frozen gases such as water vapor, carbon monoxide, and carbon dioxide, rocks and dust left from the formation of the solar system. Their orbits are highly ellipticals and are only visible when they are closer to the Sun when their proximity causes them to start melting, evidencing the popular gas and dust *cometary tails* for which comets are known for. The gas tail,[16] which can reach lengths of millions of kilometers, is made up of charged particles caused by the magnetic fields from the solar wind. The dust tail is composed of dust particles that are released from the nucleus of the comet by escaping gases.[17,18]

Fig. 2.16 Comet Hale-Bopp as imaged in March 1997. Comets usually exhibit two tails: the gas tail (blue) and the dust tail (yellow). Credit: Loke Kun Tan.

Short-period comets, also called *ecliptic* comets, have shorter orbits that are aligned near the ecliptic plane— the orbital plane of Earth around the Sun— at a distance of 50 AU from the Sun and are believed to have their origins in the Kuiper Belt. On the other hand, Long-period comets, or *isotropic* comets as they are also known, have orbits of thousands of AU from the Sun and are isotropically distributed, or in other words, their orbits are not aligned with any specific plane.

The Oort cloud is estimated to be 4.6 billion years old, and it originated after the formation of planets from the primordial protoplanetary disk. This spherical-shaped cloud is composed of trillions of objects larger than

1 kilometer and billions with diameters of 20 kilometers. These objects were formed as part of the same process that created the planets and minor planets and were much closer to the Sun than what they currently are. Most of these objects consist of water, methane, ethane, carbon monoxide, and hydrogen cyanic ices.

Gravitational interactions with young gas giants cause these objects to be scattered into extremely wide elliptical orbits. Some of them are even expelled from the Solar system into the vastness of space. The remaining ones are vaguely attached to the Sun's gravitational pull and eventually, due to gravitational interactions with nearby stars, pay us a visit in the form of long-period comets. This is the case of Halley's comet, which has a period of 76 years and was last visible from Earth in 1986. Don't forget to mark your calendars for its next visit in 2061.

Fig. 2.17 The Halley's comet photographed in 1986. It is expected to be visible again from Earth in 2061. Credit: NASA.

Comets orbiting stars other than the Sun, or *exocomets*, have been indirectly detected when the dust or gas in the extended coma has transited in front of the stellar disk.[19] However, researchers have suggested using direct imaging in the infrared part of the spectrum to detect these objects, given the large optical surface and relatively high temperature of an active cometary coma. [20] Unfortunately, current technology is incapable of distinguishing the infrared light from the comet from the infrared light from the star. Future advances in direct imaging (as we will discuss in Chapter 4) techniques may allow such a distinction.

The formation of the Moon

Structures such as the asteroid belt, the Kuiper Belt, and the Oort cloud, evidence a chaotic past where collisions and gravitational interactions between fully formed bodies and remnants of others were the norm. Astronomers refer to this stage of constant collisions as *the early heavy bombardment period,* and is believed to have occurred during the first 500 million years of the solar system's history. Throughout this period, the solar system was still forming, and planetesimal debris and leftovers from full-sized planet formation orbited the Sun. Orbits were not clear enough, and due to the gravitational influence of Jupiter and Saturn, many asteroids and comets existed. Astronomers see evidence of this era in the multiple craters present nowadays in the surfaces of worlds in the solar system; Earth was not immune to this.

A very popular hypothesis indicates that an early Earth, 4.5 billion years ago, was minding its own business when an object of one-tenth of the mass of Earth, or about the mass of Mars, impacted it. The impactor, named Theia, caused significant chunks of Earth to be thrown out to space. A fraction of all that debris stayed in orbit around the Earth and eventually became our moon. This is the description of the *big impact* or *big splat* hypothesis, introduced by Dr. William K. Hartmann and Dr. Donald R. Davis in a paper published in 1975 in the journal Icarus,[21] and currently, the leading hypothesis that explains the origin of our moon. Figure 2.18 shows an artistic illustration by Dr. William K. Hartmann himself. Hartman, a man of many talents, painted that picture in an attempt to visualize the big impact and what happened after.

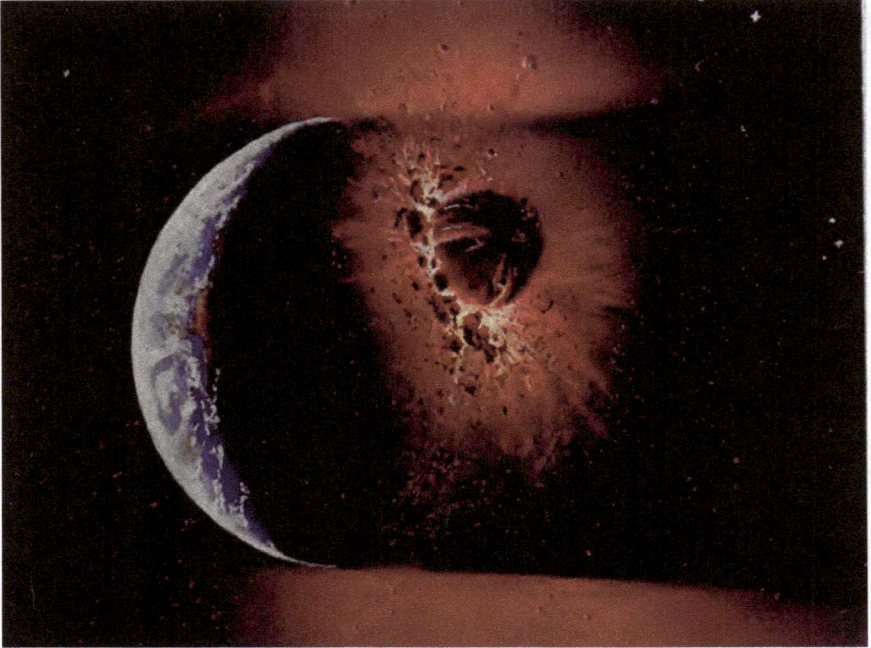

Fig. 2.18 A Mars-sized planet impacts the early Earth 4.5 billion years ago. Painting by William K. Hartmann as it appeared on the cover of Natural History Magazine in 1981.

The Big Splat hypothesis had some competitors mentioned in Dana Mackenzie's book, *The Big Splat or How Our Moon Came to Be*.[22] Dana does a very entertaining and captivating recount of the Big Splat and other once-popular hypotheses. As described in Dana's book, the leading hypothesis of the origin of the Moon during the last part of the nineteenth century is known as *the capture model*. This hypothesis was proposed by Captain Thomas Jefferson Jackson (1866-1962). Jackson's hypothesis declares that the Moon was formed elsewhere in the original solar gas cloud and "captured" when in close proximity to the Earth. Although a reasonable argument for other planets and their small moons relative to their sizes, this hypothesis fails to explain how an object as massive as the Moon, when compared to its host planet, could have been slowed down enough to be captured. In addition, the Earth and the Moon are too similar, so similar that it is quite unlikely that the Moon had formed elsewhere.

Another interesting hypothesis, the *Fission Hypothesis*, was proposed by George Darwin (1845-1912), the second son of the eminent Charles Darwin, in the nineteenth century. Darwin started working on the Fission hypothesis about 1878 and continued adding details throughout his whole life. The hypothesis argues that the Earth and the Moon were once part of a "common mass" as explained by Darwin himself in a seminar paper published in 1879. The main idea is that, in its early days, the Earth's spinning speed was much larger than it is today, causing a fragment of the planet to be expelled to space. This fragment is what we now know as the Moon. However, the Moon's orbit plane is only tilted 5.1 degrees. If the Fission Hypothesis were correct, we would expect the Moon to have a similar inclination to Earth's axis tilt (23.5 degrees). Even more, chemical composition analysis of the lunar samples brought by astronauts of the Apollo missions indicate the lack of volatiles, substances that have a low boiling point and are easily vaporized or sublimated at relatively low temperatures. These volatiles are present in surface rock on Earth, which indicate that the Moon did not form from the same material as the Earth's surface or under the same conditions.

Meteoroids, Meteors, and Meteorites

The Early Heavy Bombardment age can be behind us, but this does not mean that rocks don't continue to crash into planets' surfaces. Such rocks are named depending on the location of the object at a given point in time. Astronomers call rocks in space *Meteoroids*. Meteoroids can range in size from dust grains (0.01 millimeters) to small asteroids (10 meters). The origin of these rocks can vary from pieces of comets, asteroids, or even other celestial objects such as moons or planets. An asteroid can crash with another, break up, and form meteoroids.

Fig. 2.19 An asteroid crashes with another one and breaks up forming meteoroids. Credit: NASA/JPL-CALTECH.

When meteoroids fall into a planet's atmosphere and burn up, they are referred to as *Meteors*. These events resemble stars falling from the sky to an observer on the planet's surface. Consequently, meteors are colloquially known as "shooting stars"or "falling stars". Many people make wishes upon seeing one. I'll leave it to the reader to judge the effectiveness of such a practice.

Fig. 2.20 A meteor or shooting/falling star as it is colloquially known. Credit: Obtained from the public Internet.

Finally, if the rock does not burn completely and a part of it makes it to the planet's surface, such an object is referred to as a *meteorite*. We can see an example of a meteorite in Figure 2.21.[23]

Fig. 2.21 The Hoba meteorite. The world's largest known meteorite.

The Hoba meteorite crashed on Earth about 80,000 years ago, and given its weight of 50 tons, it still remains at the same place about 20 km west of Grootfontein, Namibia. This massive rock of 2.7x2.2 meters and a height of 1 meter, was discovered in 1920 and is estimated to be between 200 and 400 million years old. It is mostly Iron (82%) and nickel (16%) with some trace elements.

Planet Killers

Astronomers dubbed asteroids larger than 1 kilometer *Planet Killers*. These asteroids hit roughly every 600,000 years, potentially causing massive damage and terrible consequences for life on a planet. The most accepted hypothesis for the extinction of the dinosaurs is an asteroid impact.[24] Sixty-five million years ago, an asteroid, believed to have been somewhere between 10 and 15 kilometers in diameter, caused what is known as the *Chicxulub impact event*. The collision generated a blast of approximately 100 million megaton that devastated the Gulf of Mexico region, leaving behind a 180-to-200-kilometer crater. This event led to the ejection of large quantities of dust, ash, and sulfur and other aerosols into the atmosphere, effectively blocking sunlight and causing a prolonged cold winter with severe ecological cascade effects.[25]

More recently, on the early morning of the 30 of June 1908, over the basin of the Podkamennaya Tunguska River (Central Siberia), eyewitnesses saw what they described as a "fireball, bright as the sun" and then felt a powerful explosion. The event, known as *The Tunguska Event*, caused eighty million trees to be flattened and many bushes to be burned. The most plausible hypothesis of what caused such devastation is a comet or an asteroid-like meteorite that exploded at an altitude of 5 to 10 kilometers.[26]

Fig. 2.22 Felled trees caused by the powerful explosion in Central Siberia, imaged during one of the scientific expeditions in the 1920s. Credit: Leonid Kulik

The Tunguska event is so memorable that the United Nations has declared the June 30 of every year as *International Asteroid Day*[27] with the aim to "raise public awareness about the asteroid impact hazard".

A recurring joke (attributed to the American fiction writer Larry Niven) declares that "The dinosaurs became extinct because they didn't have a space program."; fortunately, we humans do.

Fig. 2.23 Due to the lack of a space program, dinosaurs could not prevent the impact of the asteroid that eventually caused their extinction. Credit: image generated by OpenAI's ChatGPT.

On November 24, 2021, NASA launched its Double Asteroid Redirection Test (DART) mission. The purpose of this mission was to investigate and demonstrate a method of asteroid deflection. Specifically, the method of deflection consisted of changing the asteroid's motion in space "through kinetic impact".[28] The targets for this mission were the asteroid Didymos and its moonlet (small moon) Dimorphos. The mission was a full success, with the impact occurring on September 26, 2022. The DART team confirmed that the impact shortened Dimorphos' orbit around Didymos by 32 minutes. This might not seem like much, but the mission was an amazing demonstration of humanity's capability to alter the course of a potential incoming asteroid.

Fig. 2.24 An artist impression of the DART mission showing the impact on the moonlet of asteroid Didymos. Credit: NASA/Johns Hopkins Applied Physics Lab

Rock collisions, although tragic in some cases, are fortunately the mechanism that caused planets to form in the first place—potentially millions of them. We began this chapter by discussing the number of planets known to us. The first planet orbiting a main sequence star other than the Sun was discovered nearly 30 years ago, and today, we are aware of more than 5,600 confirmed exoplanets. This means that planets are not just a special and rare feature of the solar system. Planets seem to be everywhere in the universe; therefore, our chances of finding life out there have increased significantly.

In the following two chapters we will explore how those planets not in the solar system are detected and the technological advancements that currently make this field one of the most exciting in astronomy.

Chapter 3
Exoplanets Detection Methods: First Part

TL;DR

Over time, our understanding has evolved from believing that the only planets in the universe were those in the solar system to recognizing that there are thousands of them. Astronomers have now detected more than 5,600 planets. Those planets, which do not orbit the Sun, are known as exoplanets. Building on the work of historical figures such as Copernicus, Kepler, and Galileo, contemporary scientists have come up with techniques or methods to detect those planets. The most successful techniques to date are the radial velocity and transit methods.

The radial velocity technique takes advantage of the Doppler effect, much like the sound of a siren from an ambulance approaching or moving away. Similarly, the wavelength and frequency of light from a star are affected when the star moves toward or away from us. For the light from a star that is moving towards us, the waves in this light will be compressed. Astronomers say then that the light from the star will get shifted towards the blue part of the light spectrum, or *blue shifted*. On the other hand, the light coming from a star moving away from us will be stretched out, resulting in the light source of the star being shifted to the red side of the light spectrum or *red shifted*. Due to gravitational effects, the presence of a planet can cause a star to move away or toward us. By analyzing the spectra of stars, the presence of planets can be inferred. This is due to the effect that planets have on the radial velocities of their host stars.

The magnitude of the change in the measured radial velocities allows astronomers to determine with some certainty the mass of the planet orbiting the star and how long it takes for it to complete a full orbit. This method also helps to determine the shape of the orbit around the planet's host star.

By far, the most successful detection method, the transit technique, takes advantage of how a planet passing in front of its host star causes a diminution in the brightness received from the light of the star. By measuring this diminution, an estimation of the planet's size can be determined. Also, by quantifying the duration of the transit, the average distance of the planet from its host star can also be calculated. Even more, a more precise way to measure the period of the planet's orbit is to measure the time between subsequent transits.

When combined with radial velocity measurements, the transit method can contribute to understanding a planet's internal composition. This is achieved by calculating the density and the planet's surface gravity.

The transit method, with an outstanding 75% of the total number of discovered planets, along with the radial velocity technique with 19%, are the obvious leaders in the exoplanet finding quest.

However, these techniques favor the discovery of large and short-period planets, as they cause larger gravitational effects in their host star and can attenuate and affect more the light collected from the stars they orbit.

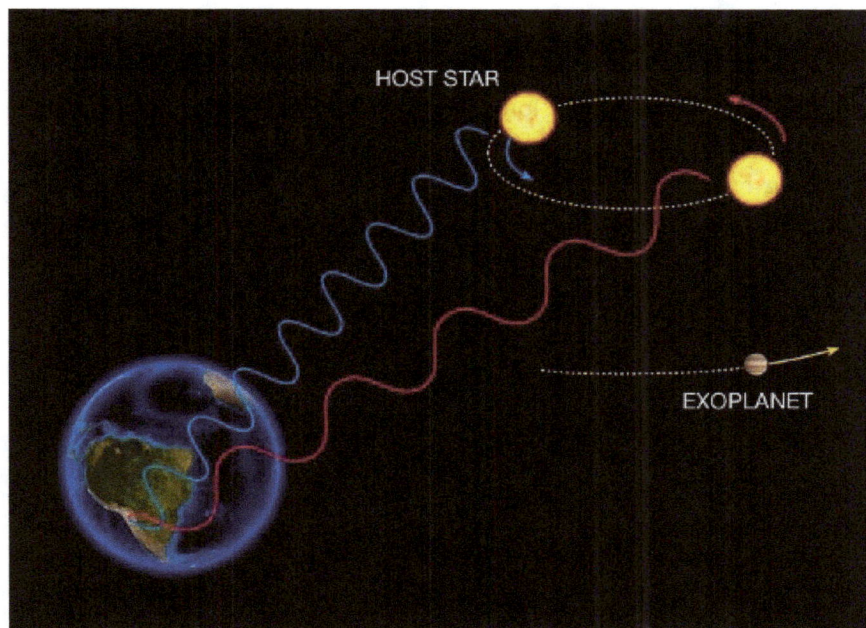

Fig. 3.1 Radial velocity method. The presence of a planet affects the starlight we receive. Credit: European Southern Observatory.

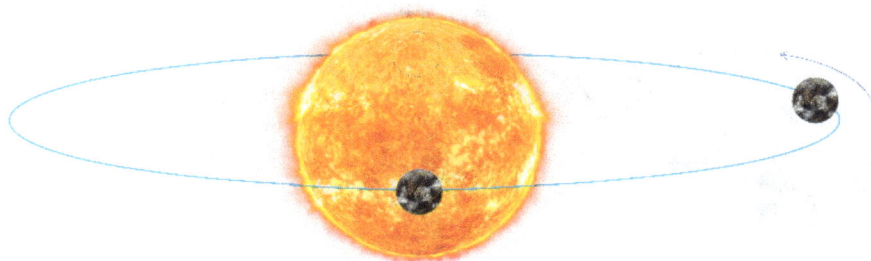

Fig. 3.2 The transit detection technique. The light of the star dims when the planet passes in front of the star. Credit: Image by the author.

Extrasolar planets (Exoplanets)

You may wonder, what is the big deal with all this recent news about detecting planets outside of the solar system? Why are people writing entire books about this? After all, we have been seeing such planets in the big and small screens for a long time now. Mostly in one of my favorite types of movies, sci-fi movies. I love sci-fi; I grew up in the mid-eighties watching reruns from the 60s of Captain Kirk and his crew landing on multiple planets and accomplishing their main goal in life: "explore new worlds, to seek out new life and new civilizations, to boldly go where no man has gone before". I still remember those words and recite them every time they are said as part of the intro of a Star Trek episode. I also watched Luke Skywalker being raised on the planet Tatooine in Star Wars (hey, who said that you have to choose one 'Star something' franchise?). Tatooine is a planet that orbits two stars (a binary star system, which is a real thing). The image of those two stars setting on the horizon of the planet is just marvelous. Can you imagine seeing that?

Fig. 3.3 The Tatooine planet. Luke Skywalker contemplates the setting of the two stars his home planet orbits in the horizon. Credit: Star Wars - A new Hope (1977).

For a five-year-old kid, whose main source of astronomy news was sci-fi TV shows and movies, I thought this was how it was; all the stars must have confirmed planets. However, as I started devouring books and magazines about astronomy, and to my surprise, I realized that the scientific community back then did not have the absolute certainty that the existence of planets outside the solar system was real. Planets orbiting stars other than the Sun were just a scientific educated guess rather than a confirmed scientific fact. I suspect that most people who also watched the same sci-fi tv shows and movies I watched were not as avid readers as myself, so they probably grew up thinking exoplanets were a proven fact. But they were not, not for a long time.

Scientists and philosophers have suspected the existence of planets outside of the solar system for a while. After all, we have a history of not being that special despite believing we are. We thought for a long time that everything was revolving around the Earth. We literally thought we were at the center of the universe. Some people still believe that about themselves (I have known way too many of them, unfortunately).

The geocentric model, proposed by Ptolemy, a citizen of the Roman Empire in the 2nd century, was a system that considered the Earth to be at the center of the universe. The *Ptolemaic system*, as it is also known, was the dominant system in Europe and the Western world until the 16th century. However, references to "other worlds" can be found as far as the 9th or 8th century BC, during the ancient Greek era. Epicurus, a famous atomist, believed in an infinite universe with infinite words.

In the Ptolemaic system, the Sun, the Moon, and all the other planets revolve around the Earth on perfectly circular orbits. The success of this model was mainly due to two reasons: first, it agreed with common sense. After all, that's what we seem to observe when we look at the sky. The Sun rises and sets, similar to the stars, the Moon, and other planets. We also don't seem to sense that the Earth moves at all. It was only logical then to think that we were static and all the other celestial bodies were orbiting around us. Second, it fits quite well with the Catholic view of our universe. Catholicism has been one of the most influential religions in human history. According to the Bible, humans are at the center of creation. It was only logical, then, to assume that everything revolves around the planet we, the special humans, inhabit.

The Heliocentric model

It was Nicolaus Copernicus (1473-1543) who dared to go against the established geocentric model of the solar system. In 1515 Copernicus affirmed that Earth was just a planet like Venus or Saturn and that all planets orbited the Sun. This model came to be known as the *heliocentrism model*, which placed the Sun at the center of the solar system. Copernicus came to this conclusion after careful observations of the brightness and movements of planets in the sky. Copernicus noticed that sometimes planets seem to be brighter. How is this possible if planets, as it was believed, followed a perfectly circular orbit around the Earth? Such a perfectly circular orbit would not mean that the brightness of planets needed to be constant? Astronomers have also noticed for centuries that planets seem to go "backward" sometimes and then forward again. This is something that you can verify by yourself. If you take a picture of a particular planet night after night and combine them, at some stage, you will see a pattern like the one captured for Mars as shown in Figure 3.4.

Fig. 3.4 Mars apparent retrograde motion. Credit: Episode 2: Wonder of the solar system, Prof. Brian Cox.

Astronomers were baffled by this behavior for centuries and tried to explain it using a lot of imagination. Particularly, considering that, according to them, planets were going around the Earth. They came up with the concept of *epicycles* or a *circle moving on another circle*. In essence, planets in this model were not going around the Earth on a simple circular orbit. Their orbit was circular, but they were "looping" from time to time, around a point circling the Earth.

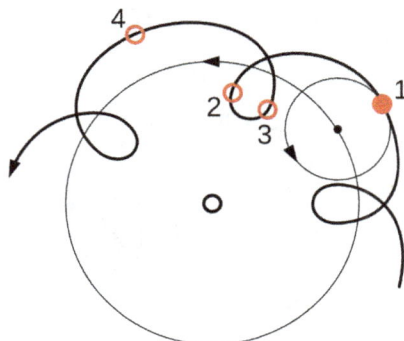

Fig. 3.5 Epicycles. A circle moving on another circle. Credit: MLWatts, Wikipedia.

Epicycles were an attempt to reconcile observations with the existing paradigm at the time that all objects in the universe orbit around the Earth. However, if a model is used where all the planets, including Earth, orbit the Sun, you don't need epicycles at all. As Kepler states later, what happens here is that planets closer to the Sun have a higher orbital speed. The further the planet is from the Sun, the slower its orbital speed. In the case of Mars and Earth, Mars moves slower than Earth as it is farther away from the Sun. This means that Earth catches up with Mars at some point giving the impression that it is static for some nights. When the Earth surpasses Mars, the planet seems to be "behind" us or going "backward". Figure 3.6 shows how the blue dot (Earth) is catching up with the red dot (Mars). For people on the surface of the blue dot, the red dot seems to go backward (points 3 and 4). This is, of course, an illusion as it only appears to be doing that.

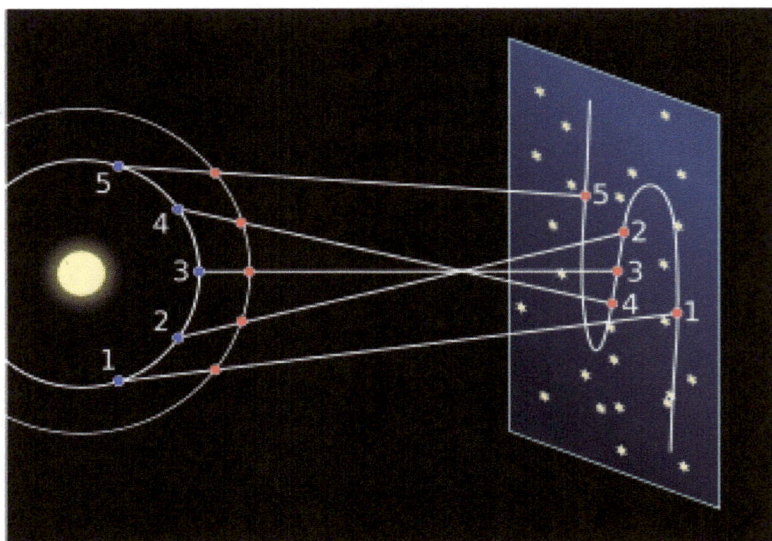

Fig. 3.6 Retrograde motion. When Earth catches up with Mars, the planet seems to be going backward. As the planets continue their paths, Mars seems again to move forward. Credit: Brian Brondel, Wikipedia.

Acknowledging that Earth and the other planets were revolving around the Sun also solves the change in the planets' brightness problem. As planets at some stages are farther or closer to Earth, their brightness also decreases, or increases, accordingly.

Okay, so the Earth is not at the center of the solar system, but for sure, the solar system is the only one out there, and the Sun is extremely special, right? Well, you already know the answer but please allow me to continue with this short history lesson. Giordano Bruno (1548-1600) an Italian philosopher, inspired by the Copernicus' heliocentric model, was the first to propose that the Sun is simply another star in the vast universe. Bruno went even further and suggested that other worlds were potentially orbiting those other stars:

"Each sun is the center of...many worlds which are distributed in as many distinct series in an infinite number of concentric and systems."[1]

Bruno was burnt alive in Rome by the Catholic church for heresy, partially for affirming the above. The final nail in the coffin for the geocentric system was placed by Galileo Galilei (1564-1642). Galilei pointed his telescope to Jupiter and observed four little bright dots going around the planet. Astronomers refer nowadays to these bright dots as the Galilean Jupiter moons Io, Europa, Callisto, and Ganymede. Galilei essentially proved that not everything was moving around the Earth.

Kepler's First Law

Now that the world could break free from the constraints imposed by the geocentric model and started embracing the new heliocentric system, people were adventurous in the pursuit of understanding more about the inner workings of the solar system. Johannes Kepler (1571-1630) played a key role in all of this. With the help of 20 years of precise observations of the Sun, Moon, and planets done by Tycho Brahe (1546-1601), he developed his famous *Kepler's three laws*. However, this was not an easy task by any means. Brahe did not just happily give Kepler his data. Kepler, who attended university to be trained as a theologian, learned about the Copernican system during his lectures and became convinced of the validity of the heliocentric model. He then moved to Prague to work as an assistant to Brahe, hoping to get access to the famous data accumulated during all those years. Brahe was not that open about sharing his observations. It has been said that Brahe's observational skills were better than his analytical skills, and he could not make too much sense of the collected data. His character was also extravagant and even jealous. Brahe did not want to share the data with Kepler, fearing that Kepler would be able to decipher the secrets of the heaven's motions, denying him all the glory and fame that such a discovery could bring. Finally, when Brahe died in 1601, Kepler was able to get his hands on all that precious data. Kepler carefully studied and analyzed the data for nearly 20 years, unveiling a big deal about the universe's secrets of motion. If Brahe had been open and cooperative with Kepler, we most certainly would know Kepler's laws nowadays as the *Brahe-Kepler* laws.

Kepler's three laws are essential to all astronomy and paramount in space navigation. From the motion of galaxies, stars, planets, and moons, the laws explain how objects revolve around other objects.

The first law deals with the shape of the orbit that a planet takes around a star. Kepler was a very religious man and had the conviction that God's creation was perfect. For centuries, starting with the ancient Greeks, the circle was considered the perfect form. Therefore, Kepler concluded the orbits of planets must be completely circular. However, regardless of how much he tried to fit the data, Brahe's observations of Mars orbit around the Sun were inconsistent with the one of a perfect circle. After massaging the data, he realized that Mars was not moving in a perfect circle, but the orbit had the shape of an *ellipse*. Ellipses are a special type of circle. They have two axes: a longer one called *major axis*, and a shorter one called *minor axis*. It happens to be that in a circle, these two axes are of equal length. Half of the distance of the major axis is referred to as the *semi-major axis*. Similarly, half of the distance of the minor axis has the name of *semi-minor axis*.

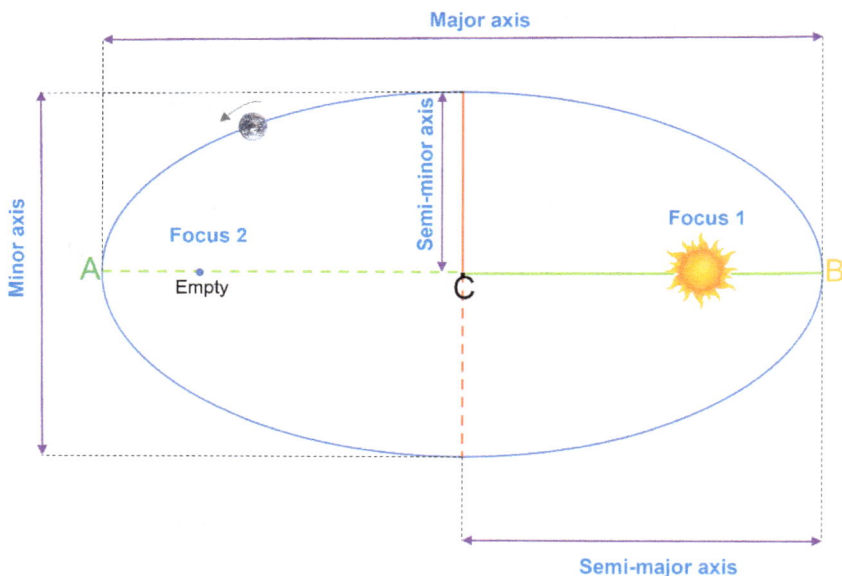

Fig. 3.7 Kepler's first law. Planets orbit around its host star describing an elliptical orbit with the star in one of the focus of the ellipse. Credit: image by the author.

Ellipses have special points called *foci*—the plural of focus— that help define their shapes. The foci are located along the major axis. Interestingly, Kepler found that for every planet in the solar system, the Sun is in one of those foci. The other focus is empty, and nothing is there. Another interesting characteristic of an ellipse is its *eccentricity*. The eccentricity is an indication of the flatness of the ellipse. It is a number that goes from zero to one. So, a circle, for instance, is not flat at all, and therefore, its eccentricity is zero. The greater the eccentricity, the flatter or more elongated the ellipse is. An ellipse with an eccentricity equal to one will be as flat as it can be and will become a segment line or in other words, a line with finite length. And this is it. This is the first law: "the orbits of all the planets around their host stars are ellipses".

Kepler's Second Law

A consequence of elliptical orbit is that planets are closer to their parent star at some stage, and farther away at other times. This explains the variation of brightness we mentioned before. For simplicity, when we read or hear about the distance of a planet to their host star, as in Earth's distance from the Sun, the value given is the length of the semi-major axis (distance C-B in the previous figure) of the planet's elliptical orbit. Note that, measured from the center of the ellipse (point C), the planet is at a distance of a semi-major length when it is at the farthest point from the star (point A), and also when it is at its closest (point B). The semi-major distance is then taken as the planet's average distance to the star.

The second law is a consequence of the elliptical shape of the orbit and is known as *the law of equal areas*. After careful analysis, Kepler came to the conclusion that a planet would sweep out equal areas in equal amounts of time during its orbit. To be able to do this, when a planet gets closer to the star, it speeds up, and when the planet gets farther from the star, it slows down.

For example, the area of region B is the same as the area of region A in Figure 3.8. The planet will take the same amount of time (t) to sweep those areas.[2]

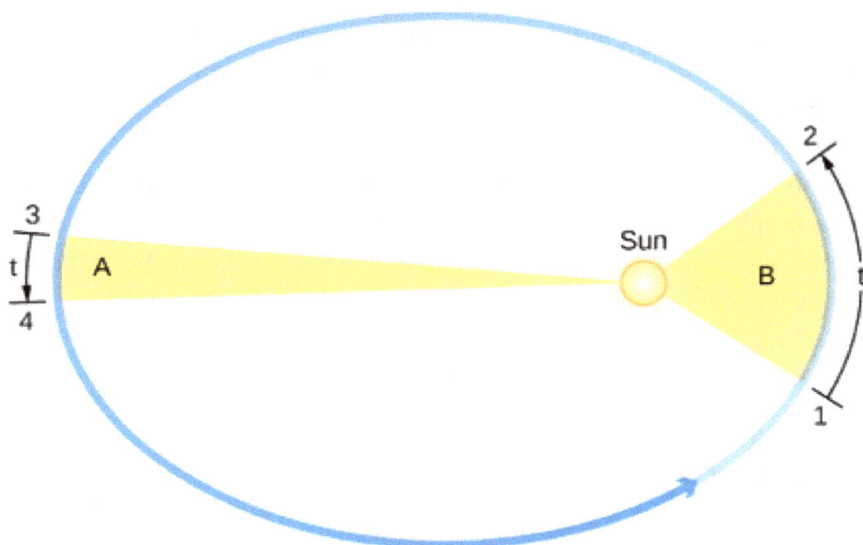

Fig. 3.8 Kepler's second law. A planet sweeps the same areas at equal intervals of time around the elliptical orbit.

Kepler's Third Law

Finally, as described by the man himself, Kepler's third law expressed the "harmony of the spheres". This law deals with mathematical patterns in the movements of the planets. Kepler found a mathematical relation between a planet's orbital period, the time the planet takes to complete a full orbit around the star, and the planet's semi-major axis (or average distance to the host star as previously discussed). Specifically, he calculated that the square of a planet's orbital period is proportional to the cube of the semi-major axis. Units are important here. This is only true when the period is measured in years and the semi-major axis in astronomical units. Let's remember that one astronomical unit (AU) is the average distance between Earth and the Sun and is approximately 150 million kilometers. With such a law, astronomers can calculate the average distance of a planet to its parent star by measuring the time the planets take to complete a full orbit and vice versa.

Newton's Laws of Motion and Gravitation

"Standing on the shoulders of giants" is a very common phrase in science. Scientific knowledge is never the result of a single person working in isolation. On the contrary, researchers and scientists rely heavily on the work of colleagues and former scientists to improve and advance human knowledge. Astronomy is not the exception. One notable example is Isaac Newton's (1643-1727) laws of motion and gravitation.

Newton is one of the most prominent scientists in history and a key figure in the scientific revolution of the 17th century. He was a mathematician, physicist, astronomer, alchemist, and theologian who devoted his entire life to unraveling the mysteries of the universe.

In 1687, Newton published one of the most important scientific texts ever: the *Philosophiæ Naturalis Principia Mathematica* (*Mathematical Principles of Natural Philosophy*). In the *principia,* Newton formulated the laws of motion and universal gravitation and utilized his mathematical formulations of gravity to derive Kepler's three laws. It is commonly said that Kepler discovered these laws empirically, while Newton provided the theoretical framework to understand why they were true.

For instance, Newton's theory of gravity explained the force that causes planets to move on elliptical orbits around their stars, as Kepler's first law established. The force of gravity also causes planets to move faster when closer to their star and slower when further away, as proposed on Kepler's second law. This behavior reflects the conservation of angular momentum, a concept we discussed in Chapter 1. Finally, Kepler's third law formulates that there is a relationship between the orbital period of a planet and the radius of its orbit. Newton's law of gravitation proved that the gravitational force between any two objects is related to the masses of the objects involved and the distances between them. Therefore, astronomers can estimate the mass of a planet with certain accuracy by knowing the mass of the star it orbits and the characteristics of the planet's period, along with observations of the star's motion caused by the planet's gravitational influence.

The Discovery of Exoplanets

Not only did Galilei, Kepler, and Newton have a significant involvement in establishing the validity of the heliocentric model and in elevating our understanding of the motion of the planets, but they were also instrumental in consolidating the idea that the Sun was just a star like any other and that there were countless stars in the universe. However, it was not until the invention of spectroscopy in the early 19th century that this was made clear. Spectroscopy breaks up the light coming from a star or the Sun into its component colors by making use of a prism. This is what we observe in a rainbow. Water particles in the atmosphere act like a prism, allowing us to see the different wavelengths (colors) that comprise the multiwavelength (white) light from the Sun.

Fig. 3.9 A rainbow appears in the skyline of Brisbane, illustrating the phenomenon when sunlight is refracted, or bent, as it passes through water droplets in the air. Captured on a cold morning following a rainfall. Credit: photo by Claudia Moreno (the author's wife).

Exoplanets Detection Methods – First Part

When astronomers verified that they observed similar patterns for starlight and sunlight, it was clear that the Sun was just another star. In particular, astronomers noted dark (absorption) features in the spectrum of different stars. Such patterns were recognized as specific elements that absorb light at specific wavelengths, effectively indicating what elements stars are made up of.

Promptly, a lot of scientists started to take Giordano Bruno seriously. Astronomers recognized that the Sun was just like any other star and realized that other stars could also host planets, just like the Sun does.

Since the mid-16th century, when Giordano Bruno made the provocative claim that the Sun was just another star, many years went by. In 1952, the forward thinking, Russian-American astronomer Otto Struve (1897-1963) proposed a set of possible techniques to practically find planets orbiting stars other than the Sun. On a two-page paper,[3] Struve laid the basis of the most successful detection techniques up to date: the radial velocity technique and the transit method.

In April 1984, the first image of a planetary disk was captured with the du Pont telescope at the Las Campanas Observatory in Chile. A disk of dust and gas was photographed around the star Beta Pictoris. This was the first time astronomers had confirmation that the planetary formation process that occurred in the solar system could have happened somewhere else in the universe.

Finally, in 1992, reality caught up with science fiction. Astronomers Aleksander Wolszczan and Dale Frail announced the discovery of two rocky planets orbiting a pulsar in the constellation Virgo.[4] In 1993, they announced the discovery of another planet orbiting a binary system composed of a pulsar and a white dwarf. Those planets were detected by using the *pulsar timing variations* technique (which will be discussed in the next chapter). As discussed in Chapter 1, pulsars' rotations are extremely regular. By measuring the arrival time of the signals coming from a pulsar over a period of time, Wolszczan and Frail identified periodic changes in a rotation that is otherwise very consistent and predictable. Those periodic changes were due to the presence of a planet or planets. Unfortunately, pulsars are rare, and even rarer is the fact that planets orbit them. Therefore, the technique developed by Wolszczan and Frail is useful only in very specific cases.

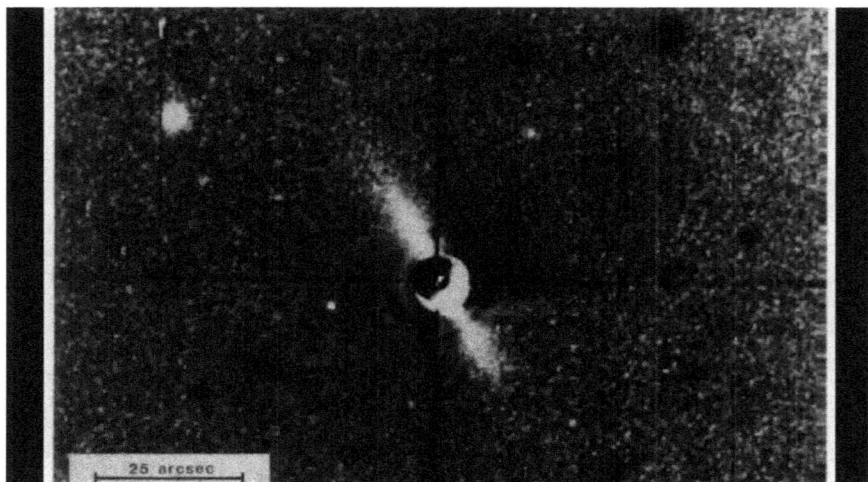

Fig. 3.10 The first image of a planetary disk around another star. Credit Bradford A. Smit, Richard J. Terrile, NASA.

The current exoplanet detection era started in 1995. Didier Queloz and Michael Mayor discovered a massive planet half the mass of Jupiter orbiting the main sequence star 51 Pegasi. 51 Pegasi is a Sun-like star located at 50.6 light-years from Earth. The fundamental difference with the discovery made by Wolszczan and Frai is that the technique radial velocity was employed to discover a planet orbiting a main sequence star. It was such a breakthrough that opened the floodgates of discovery, essentially creating a whole new field in astronomy: exoplanetary science. Queloz and Mayor shared the 2019 Nobel Prize in Physics for his work. The other recipient of the Nobel Prize that year was James Peebles for his contributions to theoretical discoveries in physical cosmology.

How Exoplanets Are Named

So, there you go. Planets orbiting other stars are real. Astronomers have detected more than five thousand of them so far.[5] Detecting those planets is not an easy feat, and requires both ingenuity and technological advances. What is much easier, however, is naming those planets. Despite how complicated it might seem, the most widely accepted and adopted naming convention is very simple.

Let's remember that in astronomy, the term *survey* refers to a systematic and comprehensive period of observation of a large portion of the sky. Discovered planets are named after the instruments employed to conduct those surveys and, in some instances, after the names of the surveys themselves. For example, The Wide-Angle Search for Planets (WASP) consortium is the most successful of the ground-based searches for exoplanets employing the transit method (which we will discuss later). A particular star being observed with this instrument will be denominated as WASP-X. Where X is the order in which the star was cataloged by position. For example, the star WASP-12 was the 12th star cataloged by the WASP instrument. Planets detected orbiting this star will be named with alphabet letters starting with 'b'. So, the first ever planet found orbiting the star WASP-12 will be called WASP-12 b. The same deal goes with the name of a survey. The planet HD 189733 b is the first planet found orbiting the star HD 189733. "HD" here stands for the "Henry Draper" catalog. A disadvantage of this naming convention is that planets are named in the order they are found rather than naming them based on their distances to their star, which can be confusing sometimes. For example, the planet Gliese-876 d, discovered in 2005, orbits closer to its parent star than the planet Gliese-876 b, which was discovered in 1998. That is baffling, at least for me. As you had expected the planet "b" to be closer to its home star than the planet "d". "Gliese" comes from the Gliese catalog compiled by the German astronomer Wilhelm Gliese in the 1960s and 70s.

Okay. So, now we know how exoplanets are named. In this chapter and the next, we will explore the techniques used to detect them. We begin by exploring the most successful techniques: the radial velocity and the transit method.

Radial Velocity Method - Finding the wobble

The radial velocity technique is why Wolszczan and Frai were awarded the Nobel Prize. This technique was used to find the first planet orbiting a main sequence star. Contrary to the pulsar timing technique that found the first actual exoplanet, the radial velocity technique is a method that paved the way to find planets orbiting a main sequence star. Main sequence stars, or "normal" stars, like the Sun, are more common compared to pulsars, which are relatively rare. Hence, opening a whole new era in human knowledge. The radial velocity method accounts for nearly 20% of all the confirmed exoplanets so far.

Before we start discussing the details of this technique, let me tell you that you and I have been deceived all our lives. Even I deceived you before at the start of this chapter. We have always been told that the planets in the solar system orbit the Sun. The same applies to planets orbiting other stars. This is not accurate. Every object has a *center of mass*. This is the exact center of all the material an object is made of. In other words, this is the point at which the mass of the object can be balanced. If the mass of an object is distributed homogeneously, the center of mass of that object would be in the middle (a ruler for instance). On the other hand, if the object's mass is concentrated towards one end of the object, the center of mass is much closer to the heavy end (a hammer, for example).

In space, objects orbiting each other also have a center of mass, which is manifested as the point around the objects orbit. This is the point at which the *barycenter*[6] (from Ancient Greek barús 'heavy' and 'center') of the objects is located. The best way to visualize this is imaging that you place the Sun and a planet in one of those seesaw toys that kids (and some adults) use to play. This seesaw is special, as it does not go up and down, but it is a rotational seesaw (don't try this at home) where both masses "orbit" around the pivot point.

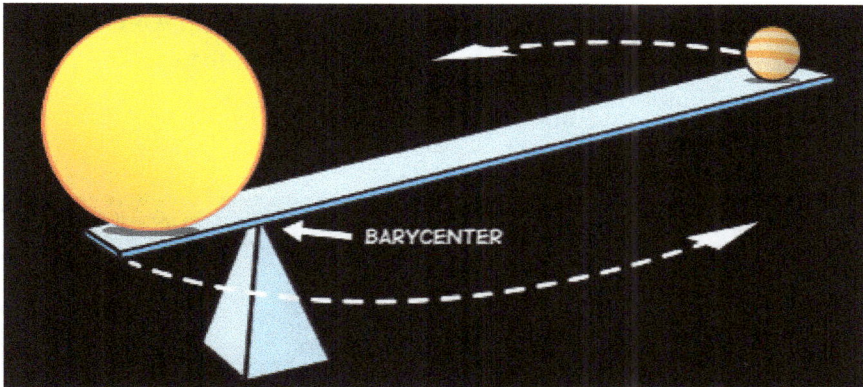

*Fig. 3.11 A planet and its host star orbit around a common center of mass
or barycenter. Credit: NASA.*

Given that the mass of the Sun is extremely large compared to any planet in the solar system, or more generally speaking, a parent star is more massive than the mass of any of its planets, the barycenter of the system will be located closer to the center of the star itself. Why is this important? Because the star and the planet, both orbit around the planet-star system barycenter. So, technically speaking, the planet does not orbit the star. It orbits the center of mass point. For less massive planets, the barycenter will be located inside the star but not quite at the center of the star itself. The more massive the planet, the farther away the center of the mass will be from the center of the star. For a planet like Jupiter, the most massive planet in the solar system, the barycenter is located about 36 million km from the center of Sun.

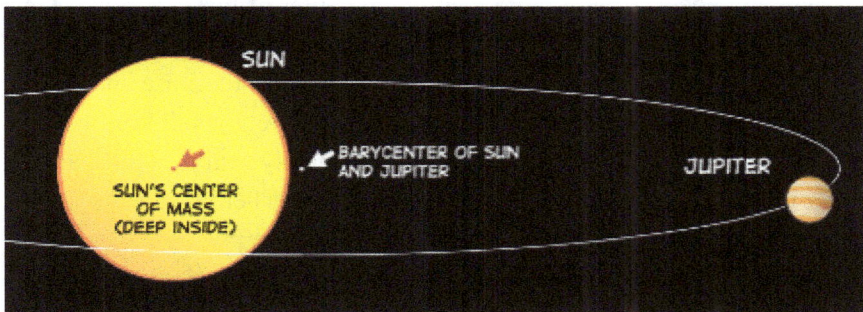

*Fig. 3.12 The Jupiter-Sun system barycenter is located at 36 million km
from the center of the Sun. Credit: NASA.*

Of course, every planet has their own special relationship with its parent star. That is, every planet is at a given distance from the star and has its own unique mass. Therefore, every planet has their own planet-star system barycenter.

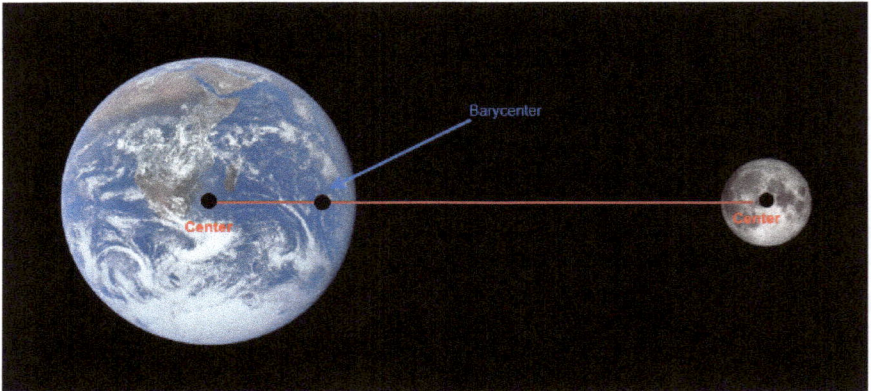

Fig. 3.13 The Earth-Moon system. The barycenter is inside Earth at 4,600 km from its center. Image not a scale. Credit: image by the author.

This, by the way, does not apply only to planets and stars. It also applies to planets and their moons, asteroids and their moons, comets, and their host star. Essentially, every pair of bodies that orbit one to the other will have their own barycenter. For instance, the Earth-Moon system's center of mass is located inside the Earth at 4,600 km from its center, which means it is inside our planet.

Now that we have established that two-body systems and multiple-body systems in general orbit around a common center of mass, it is clear that each body on a two-body system exerts a gravitational pull over the other. What this means is that a planet orbiting a star will cause the star to 'wobble'. Let's remember that planets are extremely small compared to their parent stars. The biggest planet in the solar system is Jupiter and its mass is one thousandth of the mass of the Sun. Therefore, astronomers can only detect a small wobble of the star due to the presence of a planet. The bigger the planet, the larger the wobble.

Astronomers measure thousands of spectral absorption lines per star and calculate the velocity of the star moving towards or away from us. This velocity is known as the *radial velocity*, giving its name to the detection method. Radial velocity measurement devices need to be extremely sensitive to be able to detect such small variations in velocity values. At the moment of this writing, astronomers can measure radial velocities in the range of a few meters per second (m/s) to tens of centimeters per second (cm/s). For instance, the European Space Agency's Near-Infrared Planet Searcher (NIRPS) instrument[7] is capable of measuring radial velocities of around 1 to 2 m/s; the international collaboration instrument High Accuracy Radial Velocity Planet Searcher-North (HARPS-N)[8] has a sensitivity of 40 cm/s; even more, Yale university's Extreme Precision Spectrometer (EXPRES)[9] exhibits a sensitivity of 10 cm/s.

But how can astronomers tell if the star is being pulled away from us or towards us? The answer is: the Doppler effect. We experience the Doppler effect in our daily lives. The typical example is the ambulance that is either approaching or moving away from us. As the ambulance approaches, the sound waves are compressed, resulting in a higher frequency. In the same way, as the ambulance drives away, the sound waves are stretched out behind the ambulance. Given that the distance between the stretched-out waves is further apart, the frequency decreases, producing a lower pitch.[10]

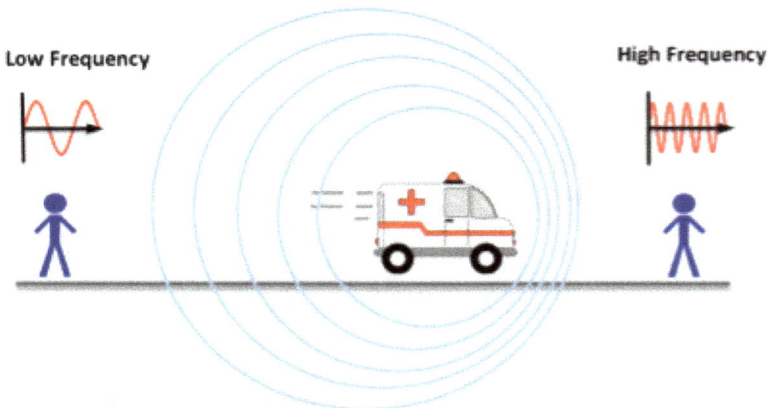

Fig. 3.14 A person will hear a higher pitch of the ambulance's siren if it is approaching. If the ambulance is going away, the person will hear a lower pitch. Credit: Online Learning College.

The same happens with light; light also behaves as a wave. Lower frequency means longer wavelength, whereas higher frequency translates to shorter wavelengths. Different wavelengths mean different colors. Particularly, the visible light goes from red (longer wavelength) to blue (lower wavelength). It is said then that the light of an object, like a star, coming towards us, due to the gravitational pull of the planet in this case, will be blue-shifted. Similarly, as the planet is completely behind the star, the planet will pull the star away from us, and the light of the star will be red-shifted. We can see this in the following figures. The movement of the star has been greatly exaggerated to illustrate the point.

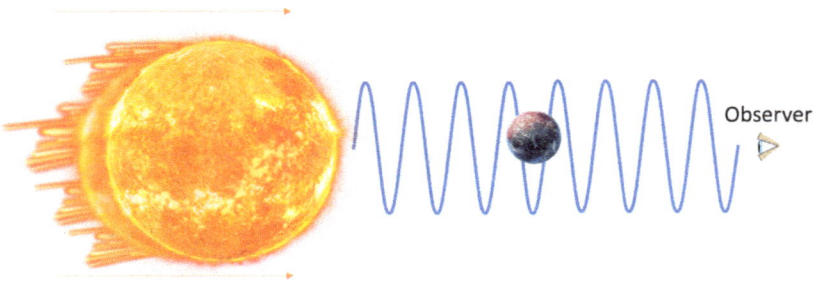

Fig. 3.15 As the star is pulled towards us by the planet, its light will experience a blue-shift. Credit: image by the author.

Fig. 3.16 When a planet pulls away its parent star the light star will be red-shifted when it reaches us. Credit: image by the author.

But what does it mean when we say that the collected light from a star has been red-shifted or blue-shifted? In the lab here on Earth, scientists have realized that every element emits photons only at certain wavelengths. When astronomers analyze the light spectrum from a celestial body, those photons show up as either emission lines or absorption lines. Similar to the example presented in Figure 3.17.

Fig. 3.17 An example of a stellar spectra.

The dark lines that you see are known as *absorption lines.* These lines are produced when the gas absorbs some of the light emitted within the interior of a star in its outer layers and, therefore, does not reach us. Spectrum lines are crucial to astronomy as they allow us to identify the atoms, elements, and molecules on distant objects, such as a star, a galaxy, or interstellar gas clouds. Here is the interesting thing. When astronomers compare the light spectrum of a star for a particular element with the spectra for the same element here on Earth, referred to as the *spectrum at rest* for that element, they can determine if the absorption lines have been red shifted or blue shifted. Hence, determining if the star is moving towards, or away from us.[11]

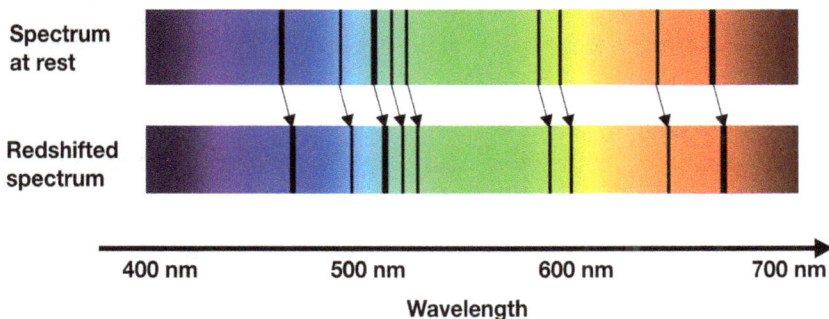

Fig. 3.18 If the absorption lines have been shifted to the red or red-shifted, the object is moving away from us. Credit: anisotropela.

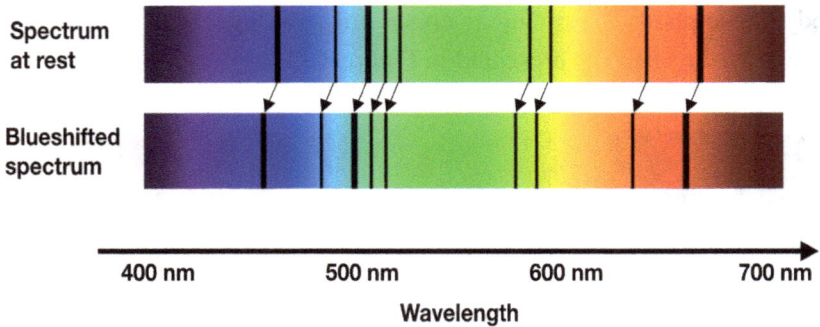

Fig. 3.19 If the absorption lines have been shifted to the blue or blue-shifted, the object is moving towards us. Credit: anisotropela.

Therefore, if astronomers observe that the light of a star moves away from us and then toward us on a periodic basis, the existence of a planet can be inferred.

The radial velocity method is one of the most successful techniques in discovering planets. It accounts for nearly 20% of the total number of planets found.[12] This method gives astronomers an indication of the mass of the planet and the time it takes to complete one cycle of revolution around its host star— the period of the planet. Radial velocity also gives researchers the eccentricity of the orbit. This parameter is extremely helpful in determining very important aspects of the planet, such as its potential habitability and formation mechanism and evolution. A planet with a high eccentricity may experience severe variations in temperature and atmospheric conditions that could limit the chances for life to thrive in such an environment.

Despite all these advantages, the radial velocity technique is not perfect. The math involved in determining the mass of a discovered planet rely on knowing the orbital inclination of the exoplanet system. The angle, i, for inclination, is the angle between the plane of the sky along our line of sight and the exoplanet's orbit.[13]

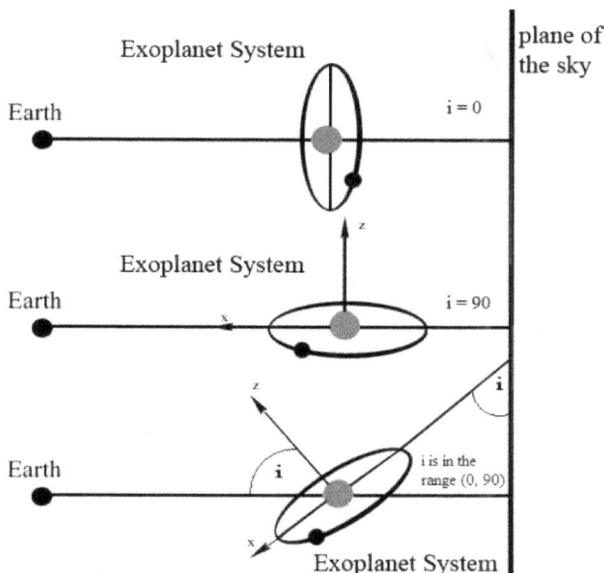

Fig. 3.20 The inclination angle i, is the angle between the plane of the exoplanet's orbit and the plane of the sky.

Unfortunately, this inclination is unknown, and as such, astronomers are only able to calculate the minimum mass of the planet. This inability to measure the planet's true mass unless the orbital inclination is known is listed as one of the main disadvantages of the radial velocity method. Determining the mass of a planet is crucial as this property is a critical criterion for distinguishing between large planets and small stars or brown dwarfs.

But the downsides of this method don't stop here. Given that stellar wobbles are typically very small, observing radial velocities far away from us is challenging.[14] Therefore, this method is often just used to examine relatively close and bright stars.

False positives are also a problem when observing multi-planet and multi-stellar systems. In such systems, the center of mass is unclear and long,[15] making it harder to calculate the radial velocity with the required precision. Misleading data can contribute to the erroneous conclusion of the presence of a planet when, in reality, such observations may be the result of complex gravitational interactions.

Another challenging problem for astronomers is that to measure the orbital period of a planet with higher accuracy, radial velocity observations of one complete orbit of the planet around its host star are ideally required. The inherent nature of the approach led to the identification of only short-term planets in the initial stages. As time has gone by, observations that have lasted for a whole decade or even longer have allowed the detection of gas giant planets in Jupiter-like orbits.[16]

The method is also biased in favor of large planets. Massive planets orbiting close to their host stars exert a stronger gravitational force on their parent stars, resulting in larger wobbles. This larger wobble produces stronger radial velocity signals, and therefore, such planets are easier to detect.

Finally, the light emitted by a star frequently fluctuates due to internal physical processes affecting the star itself. Such variations are called *stellar noise* or *stellar jitter* and also affect radial velocity signals. Star spots, for example, can produce signals that may be interpreted as a sign of radial velocity changes in the star. This can be seen as a sign of the presence of small close-in planets,[17] when, in fact, they are not there. These types of false detections are referred to as *false positives*, and astronomers, of course, want to minimize their occurrence as much as possible.

In science, like in life in general, one shoe size doesn't fit all. And this is precisely the case with exoplanet detection techniques. These techniques are not designed to be used in isolation but rather as part of the arsenal that astronomers employ to determine the different characteristics of a planet. Radial velocity is usually used in tandem with the *transit method*, my favorite technique. It is my favorite because, in my opinion, it is the easiest to understand and visualize. Let's talk about the transit method in the next section.

The Transit Method – Finding that small dip in the brightness

Imagine yourself sitting comfortably at home in your favorite chair in your living room with your lights on while you are reading the hardcover version of this amazing book. At some point, you notice this annoying insect, let's say a fly, that starts flying around. Insects, in general, are attracted to light and you fear the worst. Evidently, the irritating insect decides to start going in circles around the light bulb (it may not actually be something that insects enjoy doing, but rather an attempt to orient their backs toward the source of light).[18] Despite not following the insect's flight trajectory with your eyes, you realize that it is going in circles around the light bulb because every time it completes a full circle, the light dims a little bit. If this goes on for a while, and with regularity, you can lose your temper and try to kill the poor thing. Or, you can infer how long it takes for the insect to complete a full rotation around the bulb.

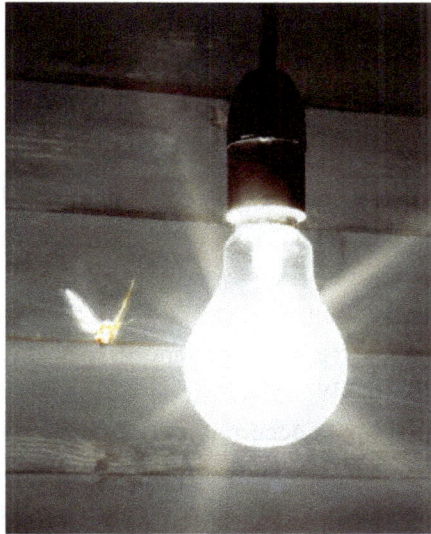

Fig. 3.21 An insect going around a lightbulb will cause a dip in the brightness of the bulb at regular intervals. Exoplanets cause a similar effect on its parent stars.

But what if the insect were smaller? That dip in brightness would be almost imperceptible. What if we substitute the fly with one of those scary big flying cockroaches? You would notice a larger dip in the light reaching you. By detecting this dip in the brightness, you could infer the presence of an insect going around the light bulb, even if you did not see the insect entering your room in the first place. You can also say that the insect is *eclipsing* the lightbulb. Well, that is literally how the transit method works.

When I was a kid back in Colombia in July 1991, I had the privilege to witness a total solar eclipse. It was an amazing experience. In the middle of the day, the Sun went completely dark. You could see flocks of birds going to sleep just before total occultation, and then you could watch those birds being so confused after the eclipse was over. What a short night, they must have thought.

But why do we get the chance to experience solar eclipses? Due to some crazy cosmic coincidence, the distance between the Moon and the Earth is 400 times smaller than between the Sun and the Earth. Or in other words, the Moon is 400 times closer to Earth than the Sun. But the size of the Moon, its diameter, is 400 times smaller than the Sun. Go figure! Due to this crazy coincidence, when the Moon is aligned just right directly in front of the Sun during a new moon phase, it completely blocks the light coming from the Sun.

Total solar eclipses have not always been experienced in the same way. Remember that the *big splat* hypothesis indicates that the Moon was created after the primordial Earth was impacted by a Mars-size object. The Moon, therefore, was much closer to Earth than what it is right now. So, total solar eclipses were not experienced back then in the same way we experience them now. When the Moon was closer, it appeared bigger in the sky, and solar eclipses occurred more often and took longer. Given that the Moon continues to slowly drift away from us, receding from Earth at a rate of approximately 3.8 centimeters per year, the Moon will appear smaller in our skies in the distant future, and total solar eclipses will become rarer and briefer, eventually ceasing altogether in approximately 600 million years.[19]

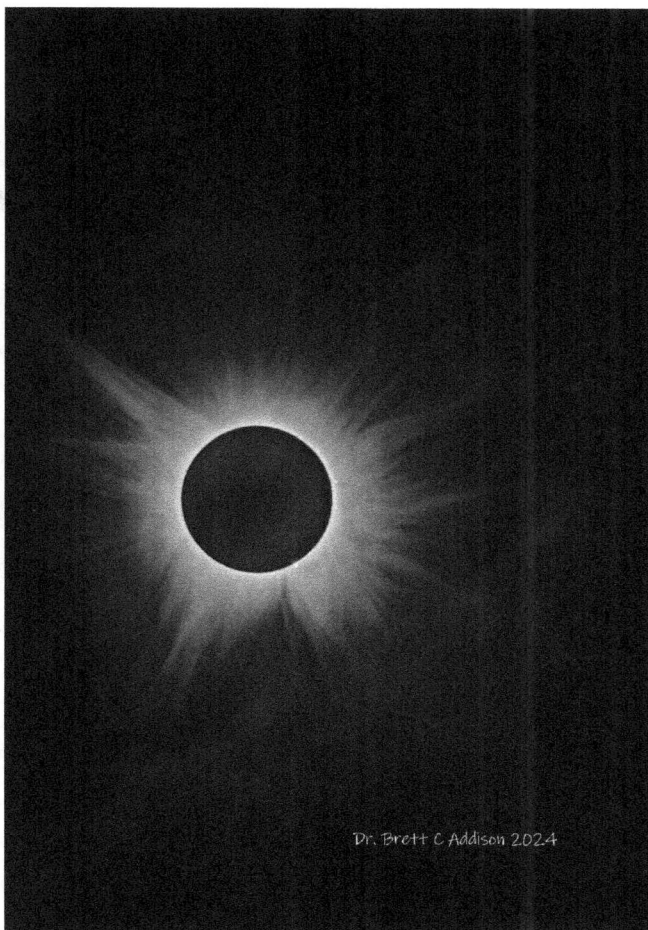

Fig. 3.22 The total solar eclipse as imaged from Texas, USA in 2024. During a total solar eclipse, the Moon completely blocks the Sun's light. Credit: Dr Brett C. Addison.

A good question to ask next is, why don't we experience total solar eclipses on every new moon? The Moon is tilted about 5 degrees with respect to the Earth's orbit around the Sun. What this means is that, from our point of view, the Moon passes either above or below the Sun, and we completely miss the resulting shadow on the Earth's surface. This 5-degree tilt also explains the other type of solar eclipses. Partial solar eclipses, for instance, occur when the Moon does not pass in exactly the same plane as the Earth's orbit around the Sun. This results in the Moon only partially blocking the Sun's light, creating a partial shadow.

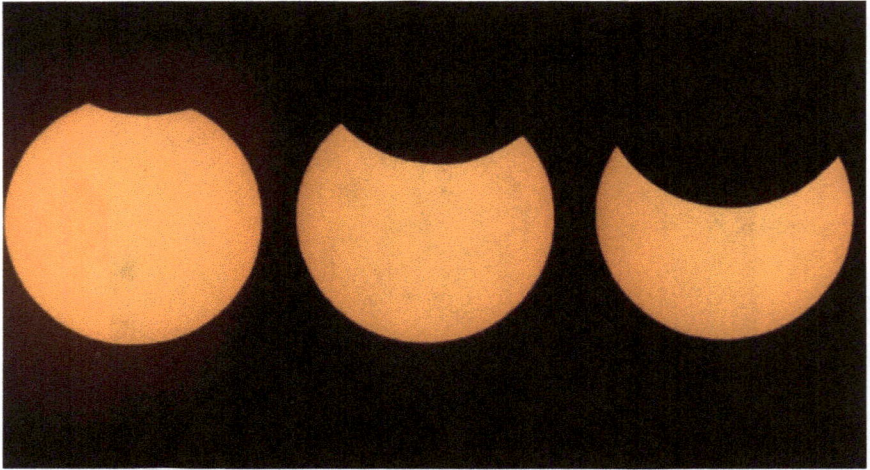

Fig. 3.23 A partial eclipse. The Sun as seen from NASA's Johnson Space Center in Houston in August 2017. Credit: NASA/Noah Moran.

Similar to what we experience here on Earth when the Moon blocks the light of the Sun, the light we receive from a star also attenuates due to the transit of a planet across its disk. Such an attenuation or reduction in brightness can be quantified using *photometry*, a technique that allows astronomers to measure the brightness of a star in an image.

To be able to quantify how much reduction in brightness the transiting planet causes, astronomers need to know the "normal" light intensity of the star. To this end, astronomers collect *light curves* of the star they are observing. A light curve is a record of the star's light intensity over time. If they are lucky, the light curve will show a consistent and periodic reduction of the received star light; the result of the planet eclipsing its home star.

From the obtained data, astronomers can determine, the *transit length*, or how long the planet eclipsed the star while passing in front of it. Astronomers also quantify the *transit depth*, which is the actual measured reduction in the star brightness. Unfortunately, a dip in the brightness of a host star caused by a transiting planet is very small. It has been estimated that for a star similar to the Sun, such a dip is in the order of 1% for planets of similar size as Jupiter and around 0.01% for planets of sizes similar to the Earth and Venus.

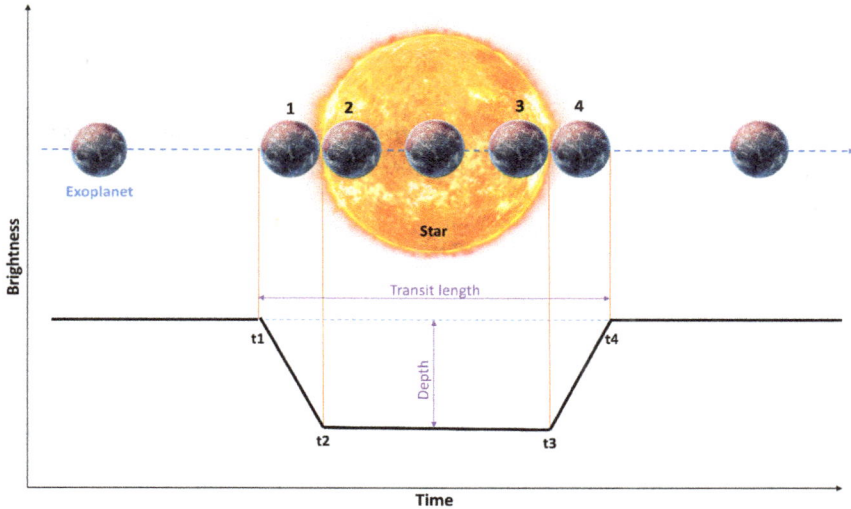

Fig. 3.24 Example of a light curve when a planet passes in front of its host star.

As observed in Figure 3.24,[20] a common terminology refers to the points of contact or phases of a transit:

First contact, which happens at t_1, or at what is known as the start of ingress, is the point at which the planet's disc is just starting to touch the outer edge of the star.

Second contact, which occurs at t_2, or at the end of ingress, refers to the point at which the entire planet has moved inside the stellar disc.

Third contact, at t_3, or at the start of egress, is the point at which the planet touches the opposite edge of the star, and,

Fourth contact, at t_4, or at the end of egress, is the point at which the planet is just outside its parent star.

The timing of these points of contact allows astronomers to reveal certain properties of the detected planet. For instance, by knowing how much variation in brightness or *change in flux*, which is the technical term, and using simple geometry, it is possible to determine the planet-to-star radius ratio more precisely. The caveat is that the size (radius) of the star needs to be known. However, astronomers have had this sorted for a while. Astronomers have produced extensive catalogues of stars sizes by analyzing the spectrum of stars, their temperatures, interferometry, or even the Moon. The mass of stars can also be determined using their spectral types.

Given the difficulty of measuring the characteristics of an exoplanet with a 100% certainty, astronomers need to make certain assumptions.

One of these assumptions is to consider the eccentricity of the planet's orbit as zero, or in other words, to consider the planet's orbit completely circular. Let's remember that the eccentricity is an indication of how flat the elliptical orbit of a planet is. An ellipse with an eccentricity of zero is effectively a circle. Doing so facilitates the calculation of the average distance (semi-major axis) of the planet to its host star. However, measuring subsequent transits of a planet enables precise measurements of a planet's orbital period, hence its distance from the star. Astronomers can indeed estimate the orbital period of a planet from a single transit based on the length of the transit, but such an estimation is not very accurate. More accurate results can be obtained if multiple transits are observed.

At the moment of this writing, 75% of all the exoplanets discovered have been detected using the transit method. In the early days, finding exoplanets with this method was primarily led by ground-based surveys such as the Wide-Angle Search for Planets (WASP),[21] and the Hungarian-made Automated Telescope (HAT).[22]

However, the field of exoplanets and the transit method saw an explosion in the number of confirmed planets when space-based telescopes entered the scene, being NASA's missions Kepler,[23] launched in 2009, and the Transiting Exoplanet Survey Satellite (TESS),[24] launched in 2018, the surveys that lead the score of the highest number of exoplanets found to date.

The Kepler mission was designed to survey more than 150,000 stars in our region of the Milky Way galaxy in the direction of the Cygnus and Lyra constellations. Its main goal was to detect Earth-size and even smaller planets in or close to the habitable zone around their star. To orientate the spacecraft to a given direction, NASA engineers designed an array of four gyroscope-like reaction wheels in the spacecraft. These wheels allowed the spacecraft to be aligned with incredible precision without needing to use fuel, which extended the mission's lifespan, which was originally expected to be four years. Unfortunately, just after three years in operation, some of these wheels started to fail; the first was in July 2012, and the second was in May 2013. Everything seemed to be lost, but NASA engineers came up with an ingenious solution. Given that the rest of the instruments aboard were still very capable of continuing with Kepler's original goal, the mission engineers came up with a very clever solution that used the pressure from the sunlight to help stabilize the spacecraft. This ingenuous trick extended the mission (known as Kepler-2, or K2 mission [25]) for another four years. Finally, after nine years of service, more than 2,700 detected planets, and having paved the way for missions like TESS, NASA announced the termination of the K2 mission in October 2018.

With the Kepler mission concluded, it was time for its successor TESS. TESS's goal is to survey 200,000 of the brightest stars close to the Sun. TESS was launched aboard a SpaceX Falcon 9 rocket in April 2018. The satellite surveys the entire sky over six years. It divides the sky into sectors measuring 24 degrees x 96 degrees.

It observes each sector for two orbits of the satellite around the Earth. This means that it devotes about 27 days on average to each sector. TESS has already helped to detect over 400 confirmed planets and counting. However, the list of possible planets (candidates) is currently more than 10,000. This means that the number of confirmed exoplanets could potentially double or even triple in the near future.

There is no doubt that the transit method has been successful. However, similar to the radial velocity technique, it also has its downsides. The most obvious one is that the transit of a planet, and therefore, an attenuation in the brightness of the host star, can only be observed when the planet's orbital plane is *edge-on*, or *nearly edge-on* along our line of sight. That is, when the observer's— an instrument's— line of sight coincides exactly or nearly so with the planet's orbit plane as illustrated in the following figures.

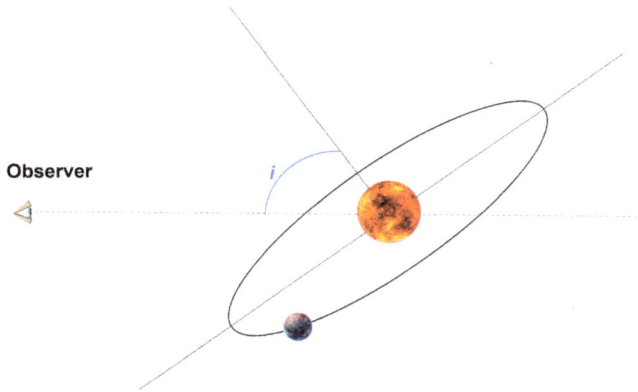

Fig. 3.25 Nearly edge-on. An inclination, i, between our line of sight and the distant exoplanet's plane orbit. Credit: image by the author.

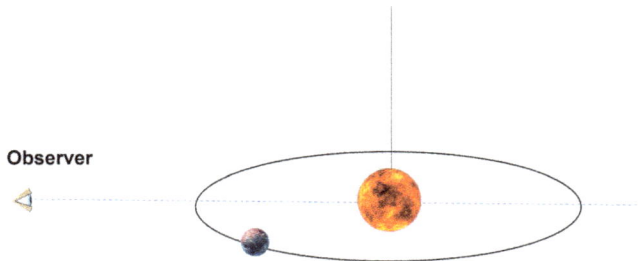

Fig. 3.26 Edge-on. The exoplanet's plane orbit is perpendicular or nearly perpendicular to our line of sight. Credit: image by the author.

Unfortunately, when the planet is *face-on*, which means an inclination of zero degrees concerning our line of sight, astronomers can't see the planet passing in front of the star and are unable to measure any change in the host's star luminosity.

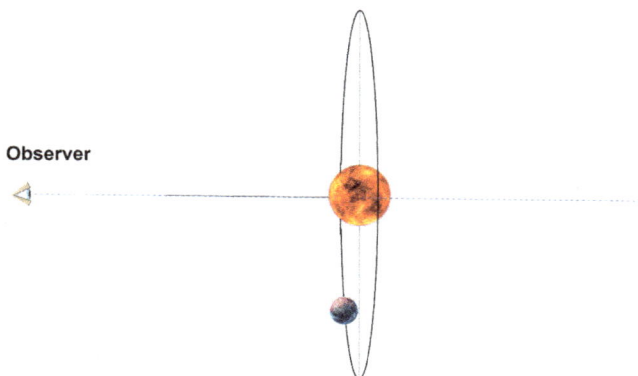

Fig. 3.27 Face-on planet. The planet's plane orbit has a zero degrees inclination with the observer's line of sight. Credit: image by the author.

The good part of this is that, in the transmit method, astronomers can calculate the value of the planet's orbital inclination by knowing the duration of the transit length and a little bit of geometry. This is done thanks to the *impact parameter*, which measures how close a planet passes to the center of its host star as it transits in front of it. Let's remember that due to the nature of the radial velocity technique, the orbital inclination of the planet's orbit was unknown.

Unfortunately, the probability of observing a transit of a given planet around a particular star is small and requires an optimum alignment between the observer and the planetary system.[26] This probability is known as the *transit probability* and is defined as the probability that a given planet will transit its parent star as viewed from Earth. This probability increases with a larger size of the parent star and decreases with larger distances of the planet from the star. For instance, for two stars of similar size, it would be more probable to observe a planet closer to its parent star than observing one on a longer orbit. On the other hand, if we have two planets located exactly at the same distance from their host star, it would be more probable to observe the planet whose parent star is larger.

The transit method is also inherently biased and favors the detection of short-period planets, which are those that orbit close to their parent star. Moreover, given that it's easier for astronomers to detect larger attenuations in the flux of a star due to the passing of a planet, the method is biased toward larger planets that orbit closer to their host star as they block more light and transit more frequently. To make things worse, the duration of a planet's transit—the time the planet is eclipsing its parent star— could only be a small fraction of its total orbital period. For example, a planet might take months or even years to complete a full orbit around its host star. However, a transit is usually in the order of only hours or days.

The transit method also suffers from false positives caused by stars belonging to the bottom of the main sequence or brown dwarfs. Such objects orbiting a larger star can be confused with giant planets as their sizes are similar.

One final disadvantage of the transit method has to do with the configuration of star systems. It is estimated that up to 85% of stars in the universe are part of a binary system.[27] These stars, also known as *eclipsing binaries,* are a particular headache for planet hunters using the transmit method. In particular, eclipsing binaries that do not completely eclipse each other, known as *grazing binaries*, can cause false positives as they can produce signals similar to those produced by an actual exoplanet.

It is also worth mentioning that false positives can occur when a background star is close to a foreground star on the sky such that the two stars land on the same detector pixel. If the background star is actually an eclipsing binary, the signal from the eclipsing binary is diluted with the light from the foreground star and the resulting light curve can mimic transiting planets—these are called background eclipsing binaries (BEBs). This has been a particularly important issue for the TESS mission as each pixel in the sky is quite large, and it is not uncommon for two or more stars to fall on the same pixel.

Leaving aside these disadvantages, we can discuss the potential of the transit method, specifically when combined with other methods. We previously discussed how detection techniques are more powerful when used together. When astronomers have radial velocity measurements of a transiting planet, they can determine a great deal of its characteristics. Radial velocities indicate the mass, and the transit method indicates the size. That is, if a planet transits, its orbital inclination is known, and the true mass of the planet can be obtained.

If you recall your physics classes back in school, you will remember that knowing a planet's size and mass will allow you to calculate its density. This is important because a planet's density can give astronomers an indication of the type of planet they are observing, allowing them to determine if a planet is gaseous or rocky. Density can also help determine the planet's surface gravity and even help understand how a planet formed. For example, a small planet with a large density, like Mercury in the solar system, may indicate that the planet has a large, iron-rich core. In the case of Mercury, the core, with a radius of 1,800 kilometers, is estimated to make up to 70% of the total planet's size.

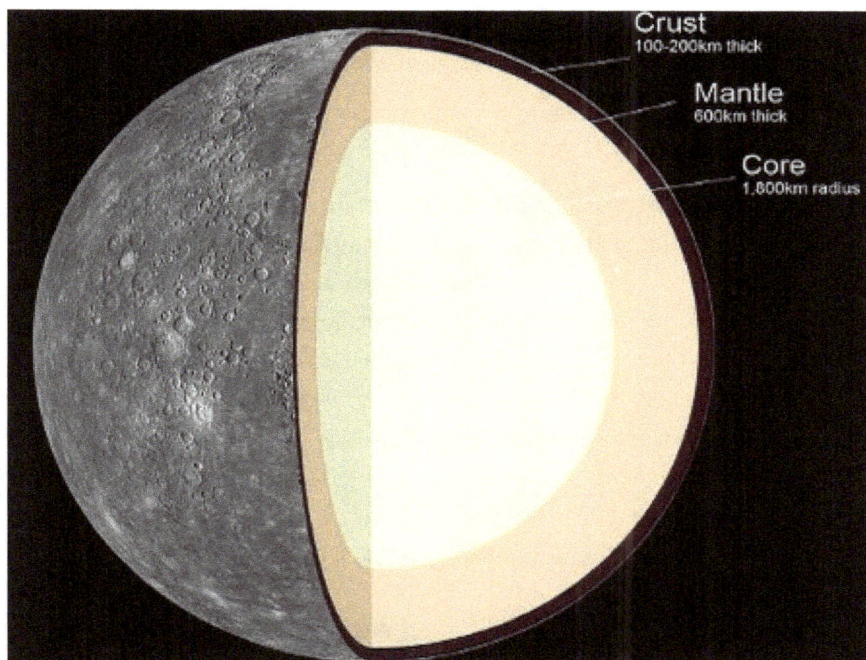

Fig. 3.28 Internal structure of Mercury. The crust is estimated to be between 100-300 km thick, the mantle around 600 km thick, and the core estimated to be 1,800 km radius. Credit: NASA/JPL.

If I haven't convinced you of the effectiveness and success of the radial velocity and the transmit techniques, let me tell you that they account for 94% of all the confirmed planets.[28]

Exoplanet Detection Technique

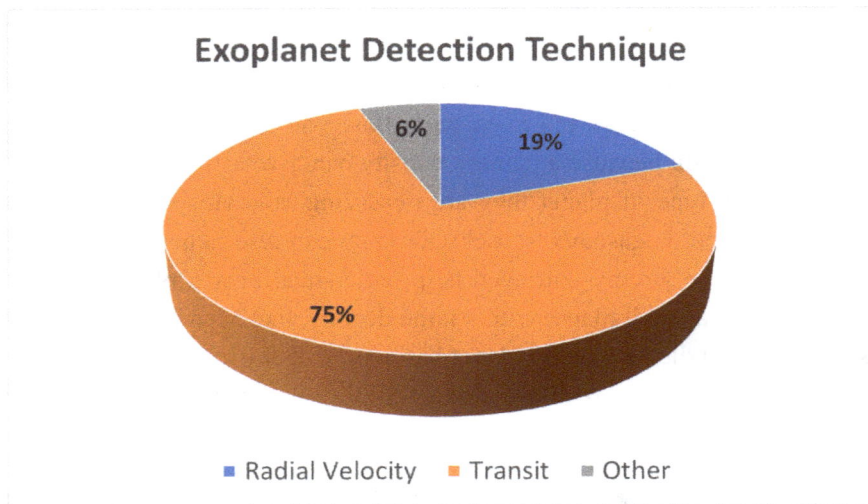

Fig. 3.29 The majority of exoplanets have been discovered using the transit and radial velocity detection methods. Credit: pie chart created using data from NASA's exoplanet archive.

But what are the techniques that account for that other 6%? We will explore them in the next chapter, and we may find that they could surpass the numbers achieved by the radial velocity and transit techniques very soon. These methods could also help astronomers find things that they have not been able to find until now. For instance, due to the nature of the methods researchers currently use, all the confirmed exoplanets so far are located within our own galaxy, the Milky Way. However, nothing makes us think that exoplanets are unique to our galaxy. For now, hypothetical planets outside of the Milky Way, the so-called *extragalactic planets*, could be found by taking advantage of the effects of gravity; let's continue our journey.

Chapter 4
Exoplanets Detection Methods: Second Part

> "Born on Krypton and raised on Earth, you had the best of both and were meant to be the bridge between two worlds."
>
> — Jar-El, Man of Steel (2013)

TL;DR

Every time a new planet is detected, a bridge is established between our own world and the one just discovered. Despite being extremely successful, the radial velocity and transit methods have several limitations. This has prompted astronomers to think hard and create other approaches to find elusive exoplanets. These other methods are worth exploring, especially since these techniques have yet to reach their critical mass. The goal of this chapter is for you to realize how bright and exciting the future of these other techniques appears.

These "other" techniques only account for about 6% of the exoplanets detected so far,[1] but regardless of their current low numbers, the potential of such methods is indisputable.

The first of these techniques is astrometry. Astrometry is a method that detects planets by precisely measuring the changes in the position of stars caused by the presence of a planet. However, given the vast distances of stars, detecting small changes in their movement due to an orbiting planet is a monumental technological challenge. Astronomers measure the movement of stars in arc-seconds. An arc-second is the 1/3600th of a degree. However, this unit is not small enough anymore. Information collected by the Global Astrometric Interferometer for Astrophysics (GAIA) has forced astronomers to introduce the milliarc-second, which is one one-thousandth of an arc-second, to measure the proper motion of the stars within its field of view. The proper motion of a star is the apparent distance on the sky that a star has moved over a certain period. While the progress of Astrometry has been slow, it is estimated that between 20,000 and 70,000 planets will be detected from GAIA data in the next 10 years.

Direct Imaging is the horse that every exoplanet fan is cheering for. It's one thing to know that there are over 5,000 planets out there, but to see a detailed picture of any of these objects would be a monumental achievement. Direct Imaging is the only method that directly detects a planet. All other techniques rely on indirect measurements to infer their presence. Obtaining such an image is not easy, as planets emit very little visible light, and the brightness of the stars they orbit are several orders of magnitude larger. Planets reflect the light from their host star accordingly to their reflection properties, or albedo. The albedo of a planet is the percentage of incoming star light reflected to space. Capturing the light reflected from a planet in the visible portion of the electromagnetic spectrum is extremely hard. But in the infrared, things look a lot better. Young planets for instance, are extremely hot and emit light in the infrared, making them the right candidates to be directly imaged. Additionally, the chances of directly imaging planets improve if they are orbiting something that is not that bright and are at a large distant orbit. Taking all of this into consideration, this technique has required amazing technological advancements. Active optics, and adaptive optics, for instance, remove the effects of the Earth's atmosphere in the collected images. Corona-graphs and starshades block the light from the parent star so that the reflected and emitted light from the planet can be analyzed. Starshades look like something straight out of an Arthur C. Clarke novel. Imagine a massive flower-like structure in front of an observatory floating in space. Science fiction, indeed!

A giant flower-like structure in space might sound like a crazy idea, but what about using a star as a magnifying glass to find exoplanets? That sounds like something made up; and yet, that is how the microlensing method works. According to Einstein's General Relativity, gravity is a space-time continuum deformation. The more massive an object is, the more deformation it causes in space-time. We witness this effect when observing the beautiful Einstein crosses and rings, which result from light being bent and magnified by a massive object between the source of light and the observer. The source of light is commonly referred to as the back-ground object, and the celestial body between the source of light and the observer, the foreground object. Exoplanet hunters use a star and any of its possible orbiting planets as a magnifier glass and search for very particular signatures in the light curves of a source of light. When a star passes in front of a background star or luminous object, and an observer, the light from the source is magnified, appearing to the observer as if the brightness of the background object has increased, temporarily reaching a peak when the involved objects are perfectly aligned. If a planet is orbiting the foreground star, astronomers will also measure a brief peak after the observed maximum brightness in the collected light curve. This tells astronomers there's a potential planet orbiting the foreground star.

Then, we have the techniques that rely on variations of otherwise regular events. Pulsar Timing Variations are one of these techniques, and the one that gave us the first exoplanet. Pulsars— rapidly rotating neutron stars, are known as galactic beacons. If the instruments used are properly posi-tioned, astronomers will see a very precise signal coming from a rotating pulsar on a regular basis. However, if a planet orbits one of these objects, it will slightly alter the orbit of the pulsar, and instruments will detect regular variations in the measured pulsation period. Thankfully, pulsars are not the only objects in the universe that produce regular signals.

Another such technique takes advantage of the fact a certain type of star regularly pulsates or varies in its brightness. One very specific type of such star is the cepheid. Cepheids are important in astronomy due to the work of the remarkable female astronomer Henrietta Swan Leavitt in the 19th Century. Henrietta found a relationship between the pulsation period of a star and its luminosity, which, in the end, allowed astronomers to refine the so-called galactic ladder and expanded our view of the universe. The Pulsation Timing Variations method observes pulsating stars and if small regular differences in their otherwise stable oscillation periods are detected, this might indicate the presence of a planet that is perturbing the orbit of the star. However, the period of oscillations is on the order of days to months, which would make it difficult to detect small changes in the pulsation timing due to the presence of planets.

But planets not only perturb stars' orbits, they also perturb other planets' orbits. This is what exoplanet hunters using Transit Timing Variations (TTV) look for. When a planet is not significantly affected by the gravitational forces of other planets in its vicinity, and it passes in front of the host star multiple times, and everything is in the right position at the right time in such a way that that those transits can be observed from Earth (I know, it is a lot of ifs), astronomers will notice that the duration and periodicity of the transit is quite regular. But, if other planets are close enough to influence the observed transiting planet in a significant way, the regularity of the transits will be slightly altered. It will be observed that the planet will sometimes transit sooner or later as its orbital period changes slightly due to the gravitational interactions with the other planets in the system. If this irregularity is periodic, this is an indication of the possible existence of a secondary planet or more.

The universe is so exotic that a planet might orbit a binary star system. Binary star systems consist of two stars orbiting each other. It is estimated that around 85% of all the stellar systems in the universe are binary or multiple-star systems, so it seems the solar system is one of the odd ones. Well, some binary star systems eclipse one another at regular intervals. The gravitational influence of a circumbinary planet, a planet that is orbiting a binary system, will disturb such a cadence and cause

Eclipse Timing Variations, which is the name of the technique used to detect these types of planets.

Coming back to planets that orbit single stars, their gravitational influence can also cause variations in the brightness of their host stars, altering the total measured brightness of a planet-star system. These Orbital Brightness Modulations are what astronomers use to infer the presence of a planet. Similar to the Moon's phases we observe from Earth, planets also experience phases, given that they reflect starlight differently as they revolve around their host star. If the observed brightness variations of the planet-star system are periodic, this can be a sign of the presence of a planet. Unfortunately, a planet is not the only reason why a star, and hence, a planet-star system, might experience variation in its brightness. Magnetic cycles in stars, like the ones that cause the sunspots in the Sun, can also regularly alter the total observed brightness of a star.

Finally, astronomers use the Disk Kinematics technique which examines the motions of gas and dust surrounding a young star. Kinematics is a branch of physics that studies the motion of objects without examining the involved forces or what is causing the motion itself. In this context, scientists analyze the movement of the gas and dust and the gaps observed in proto-stellar disks to infer the presence of a planet.

Fig. 4.1 An artist's impression of Gaia mapping the stars of the Milky Way. Credit: ESA/ATG medialab; background: ESO/S. Brunier.

Astrometry

When we discussed in the last chapter the radial velocity (RV) technique, we learned that a planet and its host star orbit around a common center of mass, or *barycenter*. We also learned that the presence of a planet can be inferred due to its gravitational effect on the parent star. In the RV technique, such a gravitational effect was manifested in the red shift or blue shift of the star's collected spectrum.

Astrometry is not exclusively used to find planets, nor was it invented for that purpose. Astrometry is an observational technique used by astronomers to precisely measure the positions and motions of stars, asteroids, planets, and even galaxies and clusters of galaxies.

Humans have been using astrometry for a long time. Our calendars, religious cycles, and holidays are based on the motions of celestial bodies such as the Sun and the Moon. Sailors have used the precise location of stars as navigation and guidance systems for centuries.

Naked-eye observations of the night sky were instrumental in determining that planets behave differently from stars, as evidenced by their movement night after night against the background sky. Hipparchus, who lived in 150 BC, discovered that the position of stars relative to the Earth changed very slowly over time by carefully observing the star Spica and comparing its position with the data collected by Timocharis 160 years earlier. Hipparchus had discovered Earth's precession. As Earth rotates, it wobbles slightly on its axis, similar to a spinning toy top. The Earth bulges at the equator due to tidal forces caused by the Sun and the Moon's gravitational pull. This causes the Earth to wobble around its axis, which is a phenomenon known as *axial precession*. The cycle of axial precession has a period of approximately 26,000 years.[2]

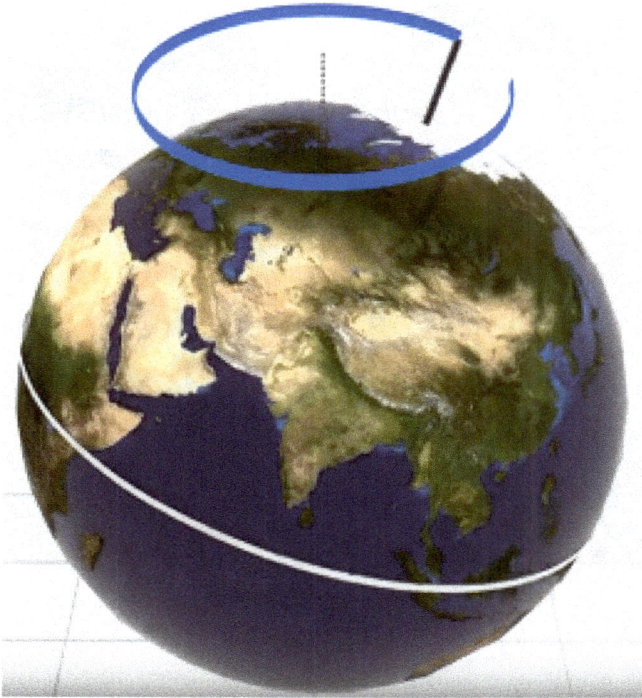

Fig. 4.2 The Earth wobbles on its own axis. The cycle of axial precession spans about 26,000 years. Credit: NASA/JPL-Caltech.

If astronomers want precise measurements of the shift of position of a star, they need to account for Earth's axial precession. The apparent shift of position of a star is known as *stellar parallax*. In other words, stars, like any other object, seem to move against a distant background when we see them from different positions. You can try this right now with a simple experiment. Find something that can serve you as a static background, a wall, for example, or whatever you can think of that does not move. Then, hold one of your fingers, the thumb is OK, in front of you and look at it with both eyes open. Then, close one of your eyes.

Fig. 4.3 Simple experiment to visualize the parallax effect.
Credit: Image created by OpenAI's DALL·E.

You will notice that the position of the thumb seems to have shifted concerning the selected background. Now, open the closed eye and close the other one. You will also notice that the position of the finger appears to have shifted again.[3]

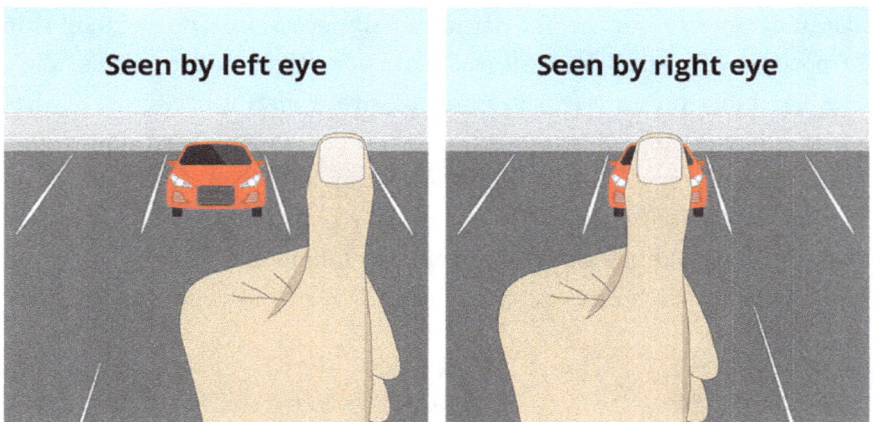

Fig. 4.4 The thumb's position seems to have been shifted depending on which eye you used to observe it.

Now, let's replicate the same experiment but with the Earth and the Sun. Earth's locations around the Sun six months apart (January and June, for instance) are similar to the positions of the eyes on our face, and a nearby star plays the role of our thumb in the previous experiment. In this analogy, the Sun would be our nose. By measuring the parallax angle and knowing that Earth's distance to the Sun is 1 AU, or 150 million kilometers, using basic trigonometry, we can determine the distance to the nearby star.

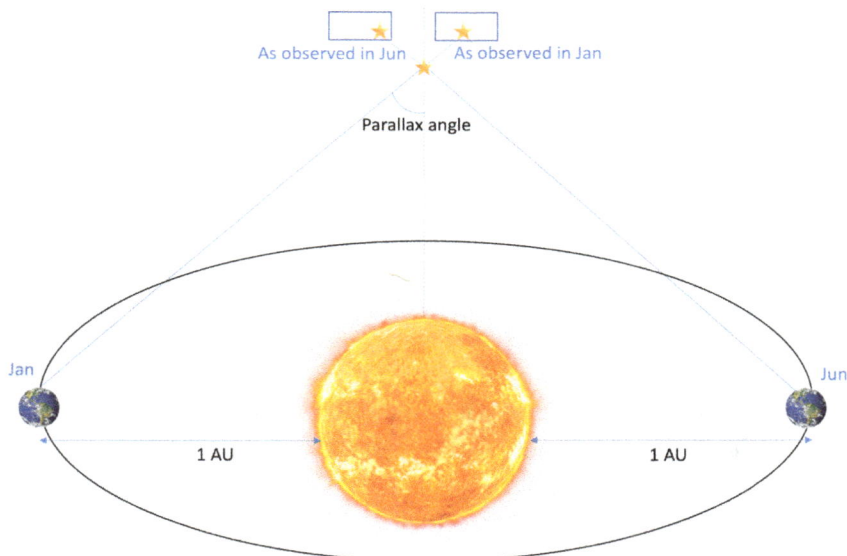

Fig. 4.5 Stellar parallax. The distance to a nearby star can be determined using Earth's position six months apart. Credit: image by the author.

Ever since Galileo pointed his telescope to Jupiter and its moons and performed the first astrometric measurements of celestial bodies using an observational instrument, astronomers realized the potential of such instruments to measure small angles.

Such a promise led to the creation of national observatories in Europe, with the Greenwich observatory, established in 1675, being the most remarkable. It was here at the Royal Observatory in Greenwich, where the first great modern star catalog, the *Historia Coelestis*, compiled by

John Flamsteed (1646-1719), was published posthumously. Flamsteed's successor at Greenwich, Edmond Halley, noticed that the bright stars, Aldebaran, Sirius, and Arcturus, had been displaced considerably from their positions when compared to previous observations. This confirmed that contrary to popular belief, the stars are not fixed. They move. The movement of a star is perpendicular to our line-of-sight, or more technically speaking, stars move *transversely*.

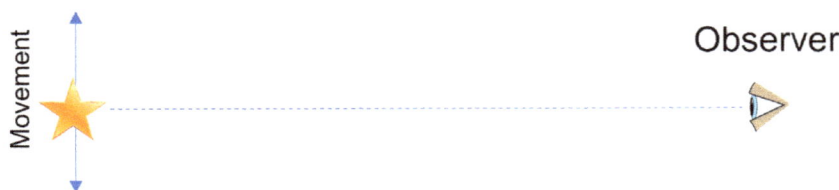

Fig. 4.6 Stars move transversely or perpendicular to our line of sight.
Credit: image by the author.

Telescopes were good at helping astronomers measure the movement of stars, but the real revolution started with the introduction of photographic plates. The plates could observe many stars in a small field of view. Soon enough, the first photographic surveys were performed. The first of these surveys, started by David Gill at the end of the nineteenth century, covered the southern sky from Cape Observatory. This type of astronomical photography used glass plates to capture images. This was the only method astronomers had to capture data for a long time. Glass plates were preferred over film because they offered greater stability and were less prone to warping or altering shape.

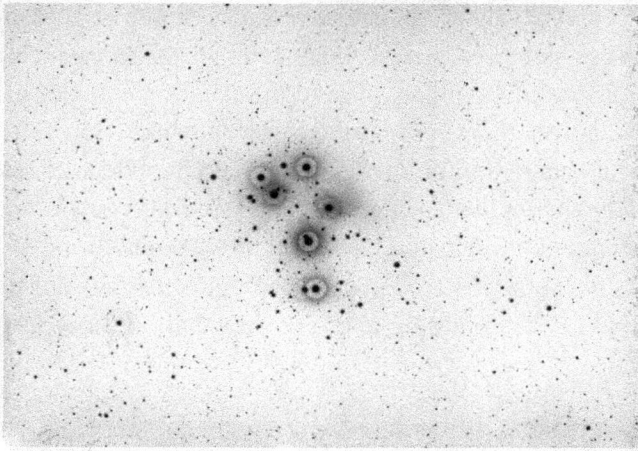

Fig. 4.7 The Pleiades captured in a photographic plate from November 24, 1951, from the Vatican Observatory archives. Credit: APPLAUSE.

Glass plates were widely used in astrometry and spectroscopy. However, the collected images still needed to be processed by hand, which was both tedious and prone to error. Then, electronics came to the rescue. The charge coupled device (CCD), the same device that older smartphones used to have in their cameras, replaced the photographic plates. CCDs are silicon chips that have the ability to convert photons into electrons. CCD's output is a digital image that can be processed to compensate for effects that reduce the quality of the captured image. These technology advancements allowed the improvement of astrometric observations.

Despite such technological improvements, ground-based observatories still suffer from several effects caused by the atmosphere that decrease the quality of the collected images. A phenomenon known as *seeing* is caused by atmospheric turbulences which cause the scattering of the light from a distant star, making it hard to determine a star's true position. To compensate for these effects, astronomers need to put observatories above the Earth's atmosphere, or in other words, they need space tele-scopes. The HIgh Precision PARallax COllecting Satellite (HIPARCOS) observatory, an ESA satellite launched in 1993, was designed specifically to act as an astrometric instrument capable of determining parallaxes.

The diameter of an object in the sky can be expressed as an angle in degrees. When doing so, astronomers refer to this measurement as the *angular diameter* of the object. Due to the size of the angles measured in astronomy, astronomers use the arc-second unit. A full circle is commonly understood to be divided into 360 equal degrees. However, it is less widely known that each degree is divided into 60 arc-minutes, and each arc-minute is further divided by 60 arc-seconds. In other words, an arc-second is $1/3600^{th}$ of a degree. For instance, the angular diameter of the Moon is 31 arc-minutes, or approximately half of a degree, as seen in Figure 4.8.

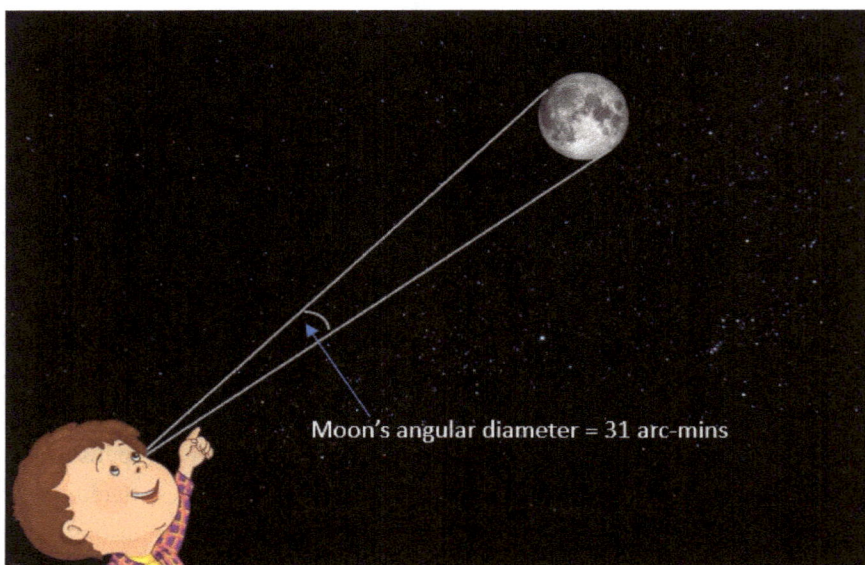

Moon's angular diameter = 31 arc-mins

Fig. 4.8 The Moon subtends an angle of 31 arc-minutes or half of a degree. Credit: image by the author.

There is a relationship between the angular diameter of an object and the object's actual diameter. The value of the angular diameter depends on the distance to the object and the diameter of the object. Therefore, a distance can be expressed in terms of an angle and vice versa. With the angle measured in degrees, arc-minutes, or arc-seconds.

If you find that the measured parallax angle is one arc-second for a given object, astronomers say that this object is located at a distance of one parsec (3.26 light-years). A more technical definition indicates that one parsec is the distance at which one astronomical unit subtends an angle of one arc-second—'subtends' is just a fancy word for 'measures'. Unfortunately, the farther away a celestial object is from us, the smaller its parallax angle, and the more difficult to be measured. Despite how good someone's eye-sight might be, humans can't detect with their eyes the stellar parallax of even the nearest stars due to the motion of the Earth around the Sun.

It is also common to express the proper motion of a star, which is the apparent distance in the sky at what a star has moved over a certain period of time, in arc-seconds or a unit derived from it. With the launch of the Hipparcos observatory in 1993, it was evident that a unit smaller than the arc-second was necessary. Hipparcos has determined the annual proper motions of stars that are 12,000 fainter than the naked-eye limit, with accuracies of 1 milliarc-second (one-thousandth of an arc-second).

An even smaller unit was needed with the launch of the Global Astrometric Interferometer for Astrophysics (GAIA) observatory in 2013. GAIA has been capable of determining the change in position of objects 4000 fainter than the naked-eye limit with accuracies of 24 microarc-seconds (one microarc-second is one millionth of an arc-second). As mentioned in the GAIA's ESA website, such an achievement "is comparable to measuring the diameter of a human hair at a distance of 1000 kilometers."

Now that we better understand how an angle can relate to a distance for observations of celestial objects, we can finally talk about astrometry in the context of exoplanets. As we have previously discussed, a planet will gravitationally affect its host star; therefore, a periodic wobble can indicate such a planet's presence. The measurements of the change in a star's radial velocity indicate the star's radial movement. Radial here refers to the component of the movement of a star that is either directed towards or away from an observer, which is used by the radial velocity technique.

On the other hand, astrometry deals with the detection of the *tangential* component in the movement of the star.

Due to the gravitation effect of an orbiting planet, a star describes an elliptical orbit (Kepler's first law strikes again!) with an average distance from the star-planet barycenter known as the astrometric signature.

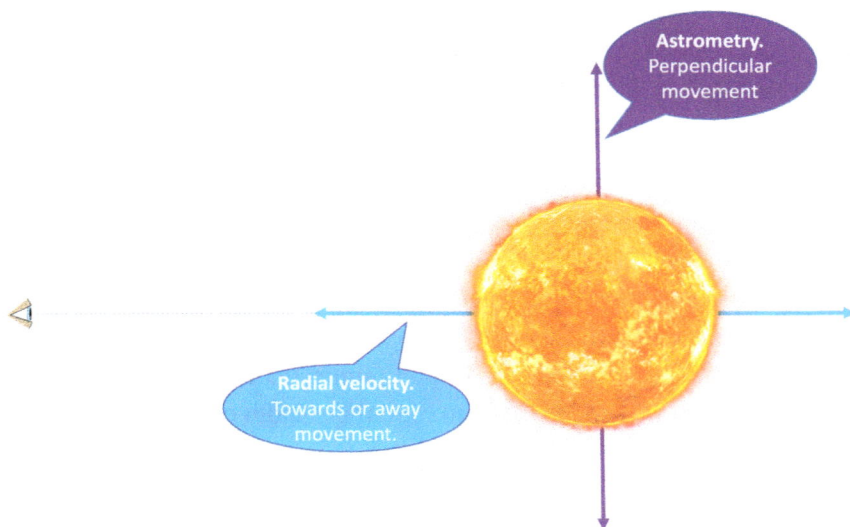

Fig. 4.9 A planet will cause a star to move on a regular basis. Astrometry deals with measuring the tangential (perpendicular) component of that movement whereas radial velocity measures the radial component. Credit: image by the author.

The magnitude of the signal for an astrometric signature depends on the mass of the planet and the distance at which the planet orbits its host star. In addition, our ability to measure such a signal also depends on the distance from Earth—or from an observer in space.

For more massive planets and those with long periods— longer than one year, the astrometric signature is easier to measure— the signature is larger. However, for a planet with similar characteristics, the astrometric signature will decrease with the distance from the Earth to the extrasolar system. This means that detecting a planet orbiting a star farther away from us is harder than detecting a planet orbiting a nearby star. In addition, measuring the wobble of a star, or practically speaking, the astrometric signature, due to the presence of a planet, is extremely hard. But, if the astrometric signature is measured with enough confidence and precision, the scientific potential is extraordinary. If the host star's mass is known, astrometric measurements can determine the value of the planet's mass without the ambiguity of radial velocity measurements.[4]

The first exoplanet found using the astrometry technique was reported in June 2013,[5] and the second one was reported in June 2022.[6] Despite the improved precision in the measurements, compared to ground-based observations provided by HIPPARCOS and GAIA, long-term observations are required to detect small changes in the positions and motions of a host star caused by a planet orbiting it. As I was writing this chapter (April 2023), a new giant gas planet orbiting a very young spectral type-A star with an estimated age between 40 and 414 million years was reported.[7] Astrometrics for the host star were collected from a combined HIPPARCOS-GAIA catalog. And that is it; three planets discovered so far using Astrometry.

GAIA's precisions are astonishing, but GAIA's main purpose is not to find exoplanets but to create the most accurate 3D map of the Milky Way. This little detail will not stop the planet hunters out there. Therefore, we can expect that many new planets will be detected with this technique. According to Dr. Perryman, the author of The Exoplanet Handbook[8] (also known as "the bible of exoplanets research"), around 20,000 and 70,000 planets will be detected from GAIA data in the next 10 years.[9] The future of Astrometry is quite promising.

Direct Imaging

All the techniques we have explored so far use indirect measurements to detect the presence of planets. In contrast, *imaging*, represents a way to observe exoplanets directly. However, stars are significantly larger and brighter than planets; even large planets are relatively small and dim in comparison to the stars they orbit. For instance, Jupiter, the largest planet in the solar system, is only one-thousandth of the size of the Sun.

Planets don't have the mechanisms to produce visible light by themselves and can only reflect a portion of the light received from their parent stars. The *albedo* of a planet is the percentage of incoming star light reflected back to space. For instance, a planet with an albedo of zero percent would absorb all the light emitted by the host star, and it would be essentially invisible to our naked eyes. On the contrary, a planet with an albedo equal to 100% would reflect the totality of the light received from the star and would be extremely bright. The Earth has an albedo of 30%, which means that about 70% of the incoming solar radiation is retained.

How much light is reflected back to space depends on the composition of the atmosphere and surface of the planet, as well as the distance from the planet to the parent star. For example, a planet with a thick and cloudy atmosphere would have a high albedo, whereas a planet with a thin atmosphere and a dark, rocky surface would have a low albedo.

The brightness of a star is much larger than that of an orbiting planet. Being able to directly distinguish the light reflected by a planet in close proximity to its parent star is challenging. Despite this, astronomers have come up with very clever ideas, and 78 planets (as per May 2024) have been discovered with this technique.

Astronomers focus on low-hanging fruit. This low-hanging fruit are hot planets. Observing a planet in the portion of the light spectrum that is visible to the human eye is challenging. However, there are also portions of the spectrum that we cannot see, but instruments can. For instance, infrared; the larger the amount of radiation from its parent star that a planet retains, the hotter it gets. A planet with these characteristics will generate its own thermal emissions and will glow in the infrared. In addition, a planet will emit radiation from thermal heat left over in its interior from formation and generate heat from the decay of radioactive isotopes.

Direct imaging astronomers often focus their attention on young planetary systems. Forming planets are still accreting material, which increases the pressure and temperature considerably at their cores, and large amounts of energy in the form of heat is released to space. It is not a coincidence, then, that the first planet detected by direct infrared imaging is orbiting a very young brown dwarf star of only 8 million years and located at a distance of 70 parsecs (228 light-years) from Earth.[10,11] The planet has a mass between 3 to 7 times the mass of Jupiter, so it is huge. It is also very far away from its host star at around 55 astronomical units! Even farther than Pluto's distance from the Sun in the solar system.

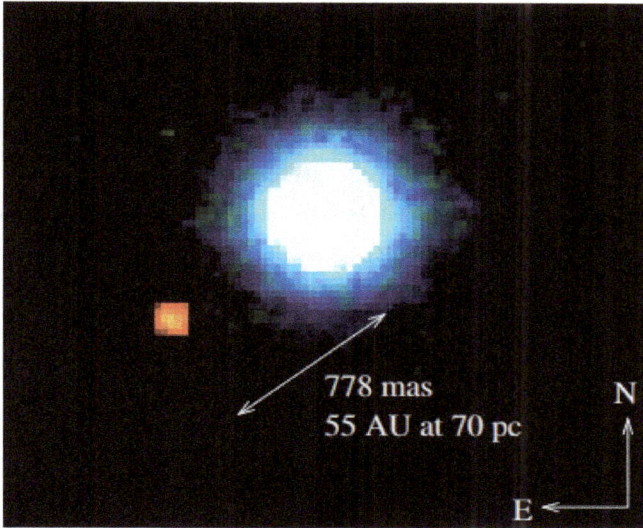

Fig. 4.10 The first ever imaged exoplanet. The little red 'blob' is a giant planet orbiting a Brown Dwarf at 228 light-years from Earth.

This first discovery is a sample of the planets that are "easier" to detect with direct imaging. Young and therefore, hot, also large, and very far away planets from dim parent stars. The reason is simple; the ratio or fraction between a planet's brightness and its parent star's brightness is very low. For Jupiter, for instance, its optical brightness is 1/1,000,000,000th the brightness of the Sun; in other words, Jupiter is 1,000 million times less bright than the Sun. This is expressed as a ratio of 10^{-9}. For Earth, on the other hand, this ratio is 10^{-10}, which means that the Earth's luminosity is 10,000 million times less than the Sun. For exoplanets, this ratio goes between 10^{-5}, or a planet being 100,000 times less bright than its host star in the infrared and 10^{-10} in the optical.

Instruments are highly sensitive to any disruptions that may impact the capture of light from exoplanets, especially considering the levels of brightness they exhibit. This is quite evident for ground-based instruments due to the effects of Earth's atmosphere. We discussed before how dry and high-attitude sites are good to remove the effects of water vapor in the atmosphere. The light coming from different sources in space traverses Earth's atmosphere in their path to the instruments on the ground. This light suffers from distortions due to turbulence and temperature fluctuations in the atmosphere, the *Seeing* effect. This effect isn't solely a concern for practitioners of direct imaging of exoplanets but rather a challenge for all ground-based observations. Observing from space is a good way of avoiding the effects of the atmosphere. Observing from space also solves the effects that Earth's gravity, temperature, and telescope alignments have on the quality of the observed images. However, sending instruments to space is difficult and expensive, especially when using large telescopes.

Thankfully, *Active Optics,* and *Adaptive Optics* (and more recently, *Extreme Adaptive Optics*) are techniques that employ computer-controlled devices to improve the quality of images by dynamically compensating for the effects of Earth's atmosphere. Astronomers use a laser guide star to create an artificial star that would act as a reference for the adaptive optic system. A laser is shot up into the atmosphere and reflected back to the telescope. The system uses the incoming light from a guide star and then corrects for turbulence effects.

Fig. 4.11 ESO's Wendelstein laser guide star system.
Artificial stars are created in the Earth's atmosphere using
a powerful laser beam. Credit: ESO/T. Kasper (AVSO).

Active Optics uses a number of electronic actuators to adjust the shape of mirrors; it monitors the image quality over time (usually seconds to minutes) to counteract the effects of environmental factors. Complementarily, Adaptive Optics reduces the effects of turbulence in the Earth's atmosphere by employing mirrors that contain voltage-responsive actuators on their surface. These actuators can deform the mirror's surface quickly (in the order of milliseconds), and continuously correct for the distortions caused by the Earth's atmosphere. The improvement in the images produced is evident, as can be seen in an image of the planet Neptune captured by the Very Large Telescope (VLT), which is located in the Atacama Desert of northern Chile.

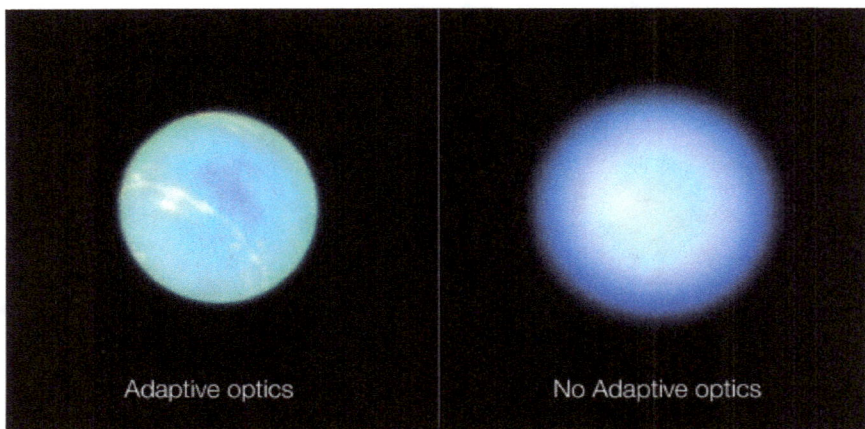

Fig. 4.12 The power of Adaptive Optics. The planet Neptune is imaged using the Very Large Telescope before and after the adaptive optics systems is employed. Credit: P. Weilbacher (AIP) and ESO.

Despite these technological advances enhancing our ability to capture and analyze the thermal emissions from exoplanets, a primary challenge persists: exoplanets are exceedingly dim compared to their host stars. Two key factors influence an instrument's ability to distinguish (in astronomers' jargon, 'resolve') the light from a planet: how much brighter the planet's light is compared to the stellar light (referred to as the contrast ratio) and how well the details of the planet can be distinguished (called spatial resolving power or angular resolution). But what if, somehow, astronomers could completely eliminate the light of the star? Well, let me tell you about *coronagraphs* and *starshades*, but first, let's talk about a very expensive toy that we briefly talked about back in Chapter 1.

The James Webb Space Telescope

All space enthusiasts held their breath on December 25, 2021. That was when the James Webb Space Telescope (JWST) was launched. JWST, a 10 billion USD instrument, is the largest optical telescope in space so far humankind has conceived. The instrument conducts infrared astronomy and it is literally helping to rewrite our understanding of the universe. Everybody has been amazed by the images delivered to the public by the

NASA press team. The image quality and the definition are incredible. One particular characteristic in the released images so far is how stars look like. Well, they look like the stars you hang on your Christmas tree, with bright spikes coming from the center.

Fig. 4.13 First full-color image released by the JWST team. Credit: NASA, ESA, CSA, STScl.

These spikes are called *diffraction spikes*. The brighter an object is, the more distinctive those spikes are. Spikes are not that prominent in objects that are not that bright, like nebulae or galaxies. Diffraction spikes are not a feature unique to the JWST. The number of spikes in JWST are eight, whereas for the Hubble Telescope, this number is four.[12]

Fig. 4.14 Image of the open cluster NGC 2660 captured by the Hubble Telescope. Four diffraction spikes are observed for stars. The bright red object is not part of the open cluster. Credit: NASA, ESA, and T. von Hippel (Embry-Riddle Aeronautical University).

Diffraction is one of the phenomena that light can experience. It occurs when incoming light hits the corner of an obstacle. The diffracted light coming from that obstacle is now considered a second source of light. The new source of light and the original source of light are then added due to the interference effect. Telescopes such as JWST and Hubble are comprised of two mirrors: the primary and the secondary mirror. The secondary reflects the light collected by the primary mirror and sends it to the eyepiece or camera.

The secondary mirror is supported by a physical structure referred to as *strut*. Diffraction spikes are then intrinsic to the secondary mirror and struts in the telescope; hence, they are unique to each individual instrument.

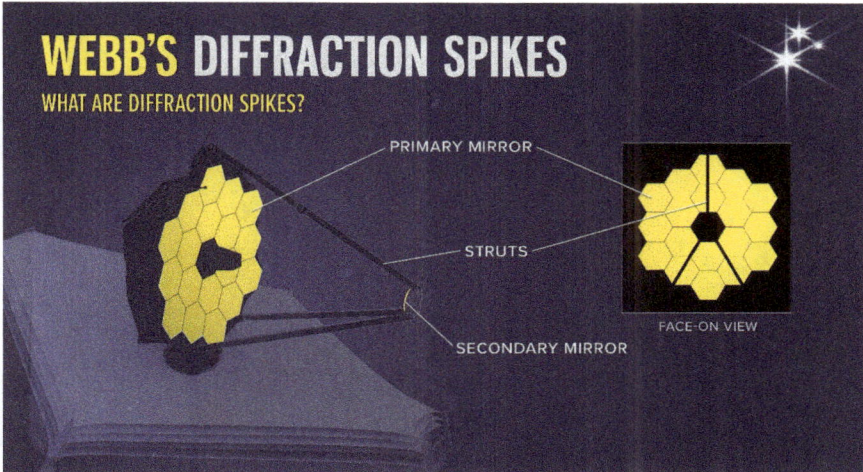

Fig. 4.15 Diffraction spikes are intrinsic to the physical features of a given instrument. Struts and secondary mirrors create the unique patterns observed in images. Credit: NASA, ESA, CSA, Leah Hustak (STScI), Joseph DePasquale (STScI).

Despite these diffraction spikes adding a nice visual touch to the images released, they are an unwanted feature for someone trying to distinguish the light coming from an exoplanet from the light of the planet's host star. To suppress diffracted starlight, astronomers have designed devices that are placed inside telescopes. These devices are known as *coronographs.*

Coronagraphs

Coronagraphs, originally developed by the French astronomer Bernard Lyot (1897-1952) in 1939, were designed to suppress the light from the Sun, allowing researchers to study the coronae of the Sun. Coronagraphs employ digital signal processing to create a copy of the received starlight to create a destructive interference pattern, effectively removing the light of the star.

In June 2022, the JWST was used to directly image the planet HIP 65426 b, a giant planet with 9 times the mass of Jupiter that orbits an A-type star. HIP 65426 b is more than 10,000 times fainter than its parent star.

JWST used different filters to image the planet at different wavelengths and used a coronagraph to block the light of the star.[13,14]

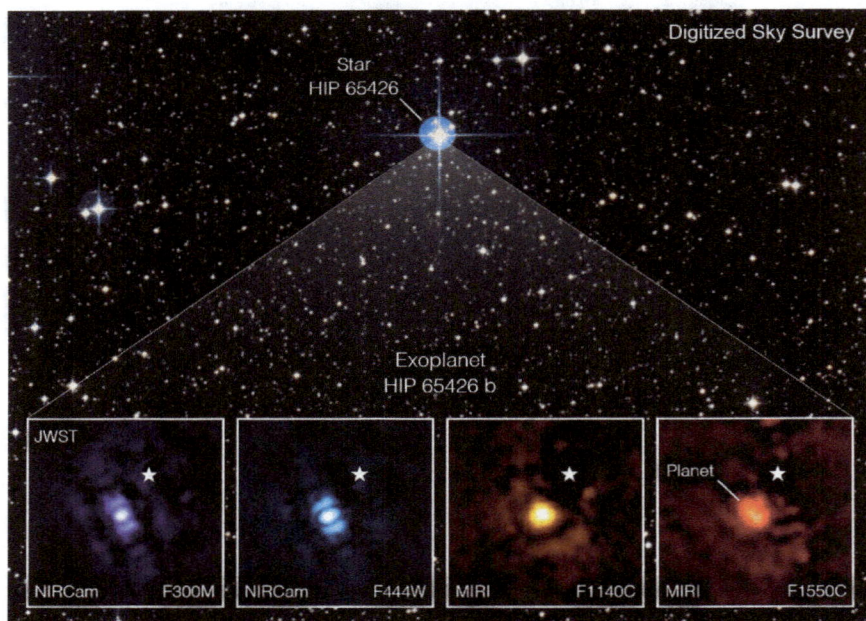

Fig. 4.16 The exoplanet HIP 65426b is observed with the JWST in different bands of infrared light. The coronagraphs or masks block out the parent star's light. The small white star marks the location of the star HIP 65426.

These results from the JWST are impressive, but the direct imaging technique is just getting started. The still-in-development NASA's state-of-the-art Nancy Grace Roman Space Telescope, (formerly the Wide-Field Infrared Survey Telescope, WFIRST),[15] or Roman for short, employs a very advanced and complex array of actuators that move like pistons, changing the shape of two flexible mirrors inside the instrument. This design allows the coronagraph to adapt to a particular starlight pattern. Additional image processing software improves and enhances the light received from a planet. Scheduled to launch in July 2026, Roman, which will observe in the visible and infrared part of the electromagnetic spectrum, will be able to image Earth-like planets 10 billion times dimmer than their host stars. Roman can also *characterize* those planets using spectroscopy, allowing scientists to study and measure the physical and atmospheric properties of exoplanets, including their sizes, masses, temperatures, and atmospheric compositions. This means that astronomers could identify elements in the atmospheres of these Earth-like planets.

Coronagraphs are a great resource to suppress the light of a parent star. However, given that coronagraphs are inside the instruments, starlight still reaches the telescopes where it can scatter and obscure the very dim planet.

Starshades

To prevent the light from a star from even reaching the telescope, an external artifact known as *starshade*, or *occulter*, has been proposed. The occulter is placed between the telescope and the star and is usually envisioned as being in space. A very dark and highly controlled shadow is produced and the telescope, which could be either on the ground or in space, is placed in a location within the area where the shadow is casted.

The best shape for the occulter is not spherical as one might think. Suggested by Lyman Spitzer at Princeton in his visionary paper *The Beginnings and Future of Space Astronomy* in 1962,[16] a starshade or occulter could be used to image exoplanets.[17]

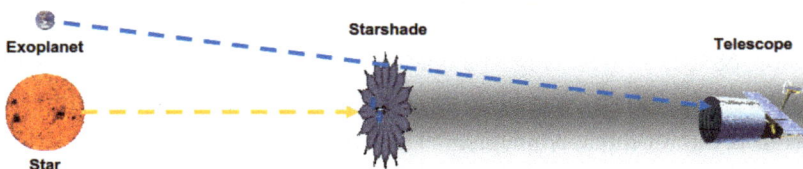

Fig. 4.17 A starshade in space prevents the light of a parent star from reaching the space telescope, allowing the imaging of an orbiting exoplanet.

Spitzer realized that a circular disk would not be enough to image an Earth-like planet as such a shape would suffer from high levels of diffraction at its edges. The idea of using an occulter to image planets was revived by G.R Woodcock of the Goddard Space Flight Center in 1974, who suggested the use of an *apodized* starshade. The term

'Apodized' comes from *Apodization functions* in math. These functions describe the gradual reduction of a signal or a function towards zero at its edges. In astronomy, apodized makes reference to a type of mask that consists of an opaque disk surrounded by shaped petals. Such a design gradually decreases the intensity of light in the direction of the edges, effectively reducing the amount of light scattered or diffracted around the object that is occulted, in this case, a star. This significantly improves the clarity and quality of the collected image.

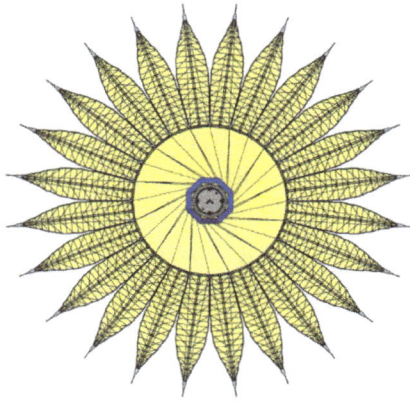

Fig. 4.18 Example of an apodized occulter. The design consists of a central dark circular structure with petal-like structures around it. This pattern reduces the diffraction effects around the edges of the starshade. Credit: NASA.

Mission concepts involving starshades have been proposed for a while now. NASA's Exo-S Starshade Probe-Class Exoplanet Direct Imaging mission is one of those concepts.[18] The mission recommends an apodized starshade flying in formation with a telescope in space. The telescope is nothing out of the ordinary. It is a 1.1-meter telescope similar in size to the telescopes you can buy at your local astronomy shop. On the other hand, the starshade is enormous, and its design depends on the launch configuration of the mission. One option is to launch the telescope and the starshade together on the same rocket. In this scenario, the starshade would consist of a 16-meter inner disk and 22 petals, each 7 meters long. Alternatively, the telescope could be sent first, followed by the starshade at a later time. This allows for a larger starshade, with a 20-meter diameter inner disk and 28 petals, each 7 meters long. Exo-S ambitions are very clear; it intends to discover new planets from Earth size to giant planets. This is theoretically possible as it could see planets that are 100 billion times (10^{-11}) dimmer than their parent stars. A second objective is to measure the spectra of newly discovered planets and known planets to identify the components of the atmospheres of those exoplanets. In addition, the mission aims to characterize planetary systems, particularly focusing on improving our understanding of the dust that surrounds host stars. This is crucial because acquiring such a piece of knowledge will enhance our comprehension of asteroids and comets.

The main disadvantage of starshades or occulters compared to coronagraphs is their cost. Launching objects into space is expensive, especially if you have to fly both the telescope and the starshade. The larger the telescope's diameter, the more expensive the whole mission gets. For that reason, other alternatives have been proposed to reduce costs. Markus Janson from the Stockholm University and his team have proposed a solution that only requires the occulter to be sent to space while taking advantage of large telescopes on the ground. In their proposal,[19] they conceive an apodized occulter in an orbit that changes to maximize the time its shadow is cast for a long time to a given location on the ground. In that location, Markus and his team propose to have a big telescope, specifically the one Europe is building at the moment. This big telescope is the European Extremely Large Telescope, or E-ELT, a 40-meter telescope scheduled to be delivered in 2025 and located at Cerro Armazones in Chile. The E-ELT will be able to collect 13 times more light than any of the existing optical telescopes of today and will be fully equipped with adaptive optics capabilities. Markus and his team's idea obviously reduces costs by not having the telescope in orbit. However, there are many challenges to such a mission. Optimal orbits are extremely important as astronomers want to maximize the number of planetary systems they want to observe and the duration of the shadow in the selected location. Altering or maintaining those orbits requires fuel that will need to be transported along with the occulter at the time of launch. Bringing more fuel means higher costs along with a larger rocket to accommodate the extra payload.

Regardless of all the limitations and potential issues, direct imaging is the method that sparks the imagination of all of us exoplanet enthusiasts. The possibility of observing an Earth-like planet in the habitable zone of a Sun-like star is definitely something worth cheering up for. However, due to the nature of the technique, direct imaging, at least in the short term, favors the discovery and characterization of large and close-by planets. For a technique more suited to finding smaller and farther away planets, we will need to go back to those techniques that detect exoplanets indirectly. Specifically, the *microlensing* technique, which we will discuss next. Regardless, we can all agree that the future for direct imaging looks very bright (pun intended).

Microlensing

I believe that if Einstein were alive, the microlensing exoplanet detection technique would likely be his favorite. Einstein's General Relativity, proposed in 1915, defined gravity as a deformation or curvature of the space-time caused by massive objects. In Chapter 1, while discussing the nature of black holes, we discussed how light is bent due to the presence of mass. If we have a massive object (foreground object) between the observer and a distant source of light (background image), the gravity of the foreground object might cause the distortion of the light coming from the background object. For this reason, the foreground object is also known as the lensing object. How that distortion is presented to the observer depends on the alignment of the source, the mass of the object in the middle, and the observer. For instance, Figure 4.19 shows an *Einstein cross*, or four copies of the background object—the quasar in this example— surrounding the galaxy, which acts as the foreground object.

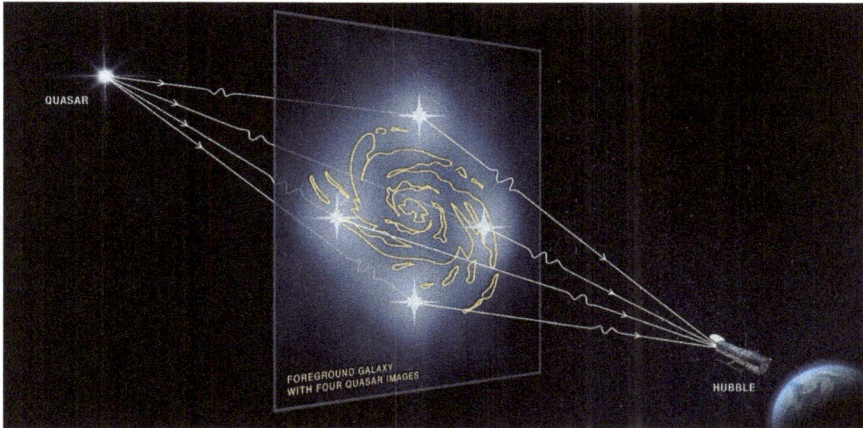

Fig. 4.19 An Einstein Cross or four copies of the background image. When a massive object is located between a distant source of light and the observer, the light from the source gets distorted due to the large gravitational field of the foreground object. Credit: NASA, ESA, and D. Player (STScI).

The distinctive shape of an Einstein Cross is due to the asymmetric gravitational lensing caused by the asymmetric mass distribution of the foreground object. However, if the gravitational lensing is symmetric, and the geometrical alignment is just right, the resulting image is an *Einstein Ring*. In an Einstein Ring, the light from the source is distorted and focused on a ring-like structure around the foreground or lensing object. Einstein predicted the existence of these structures, but he was skeptical that we would ever be able to observe one.

Einstein wrote in 1936:

"Of course, there is no hope of observing this phenomenon directly. First, we shall scarcely ever approach closely enough to such a central line."

But astronomers do observe them. The first image of an Einstein Ring was captured by the Very Large Array (VLA) in 1987.[20] It was an image taken in the radio portion of the spectrum at a frequency of 1.49 Gigahertz (GHz).

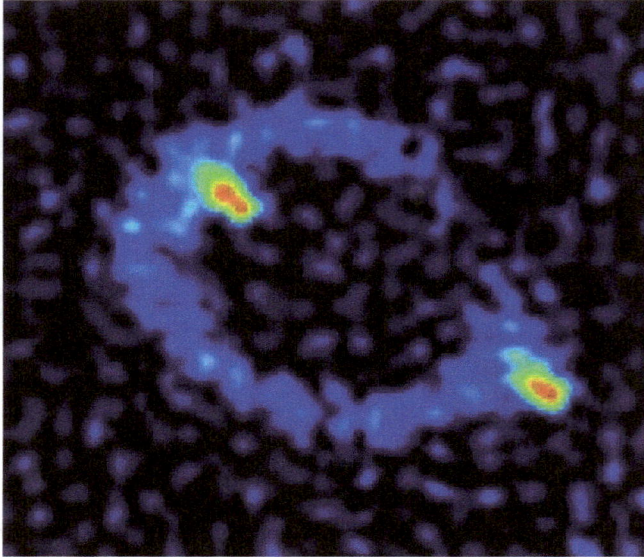

Fig. 4.20 First ever photographed Einstein Ring. Credit:
VLA.

Twenty years later, in 2007, the Hubble Space Telescope, observing in the visible and infrared ranges, produced an even more beautiful and detailed image of an Einstein Ring around the galaxy LRG 3-757.

Fig. 4.21 An Einstein ring surrounding the galaxy LRG 3-757. This image was produced by the Hubble telescope in 2007. Credit: ESA/Hubble & NASA.

In the exoplanet context, the foreground object also referred to as the *lensing object* is a star. As the lensing star passes in front of a background star—the lensed object, it acts as a magnifying glass, creating a temporary increase in the brightness of the background star. Even more, a planet orbiting the lensing star will create a distinctive feature, or microlensing signature, in the collected light curve of the lensed star.

This method can be described as the opposite of the transit method. As discussed in Chapter 3, this method detects and characterizes exoplanets by measuring the decrease in brightness of a host star caused by the presence of an orbiting planet. In the microlensing method, astronomers detect and characterize exoplanets by measuring how much brighter a background star gets when a star-planet system passes in front of it.

If the lensing star hosts no planets, the light curve for the background star will typically exhibit a bell-shaped profile. The peak of this light curve occurs at the point when the lensing star passes in front of the lensed star.

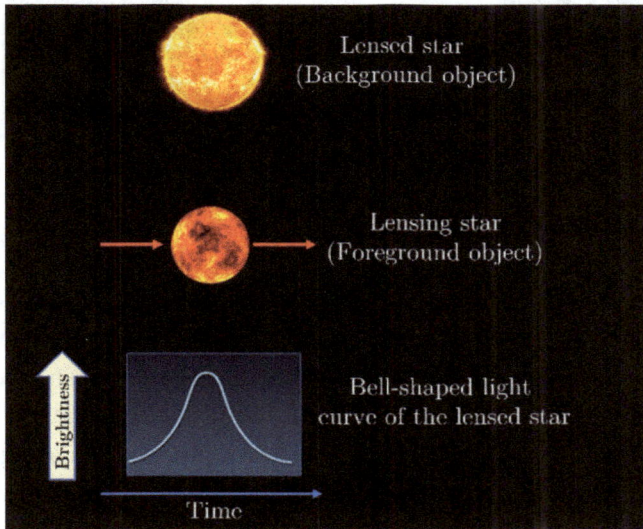

Fig. 4.22 An observer will see how the luminosity of the background star starts to increase when the foreground star passes in front. The peak of brightness occurs when the stars are perfectly aligned. Credit: NASA, ESA, and K. Sahu (STScI).

If the lensing star does have an orbiting planet, the planet will also briefly gravitationally magnify the light of the background star, increasing its brightness independently. This will produce a very distinctive feature in the observed light curve of the lensed star.

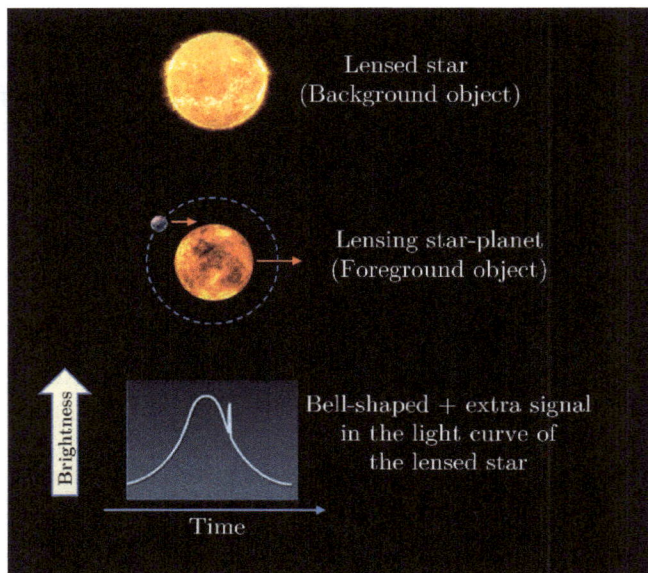

Fig. 4.23 A planet orbiting the lensing star will create a very particular signal in the light curve of the background star. Credit: NASA, ESA, and K. Sahu (STScI).

That extra distinctive signal allows astronomers to characterize the orbiting planet. By analyzing the duration and the shape of the magnification in brightness caused by the planet, its mass, the distance to its host star, and orbital distance can be inferred. For instance, the larger the planet's mass, the larger the gravitational effect it produces, causing a longer duration of the brightness event. In addition, the shape of the obtained magnified light curve helps to determine the distance and orientation of the planet's orbit.

The first exoplanet detected using the microlensing method was a planet orbiting the low-mass star, OGLE-2003-BLG-235L, located about 26,000 light years away from data collected from the Hubble Telescope. The planet, OGLE-2003-BLG-235L b, discovered in 2004[21], is nearly three times the mass of Jupiter and orbits its host star at 4.3 AU of distance. It was found using the instrument Optical Gravitational Lensing Experiment (OGLE) at the University of Warsaw, Poland.

Unfortunately, the alignments between a background star and a lensing star-planet system are rare and unpredictable, which is the main disadvantage of this method. This means that follow-ups of such events are very unlikely as they only happen once. If possible, follow-ups will need to be done using another technique. The way to overcome this rareness and unpredictability is to observe as many background stars as possible and train computer programs to detect the pattern in brightness caused by a lensing event generated by a star-planet system. This method is followed by surveys such as OGLE,[22] the Probing Lensing Anomalies NETwork (PLANET),[23] and the Korean Microlensing Telescope Network (KMTNet).[24] These surveys collect data from thousands of background stars searching for microlensing events. Although not all identified microlensing events are attributable to star-planet systems, scientists have developed methods to distinguish them from the collected data. Despite the unpredictable nature of the microlensing method, it has led to the discovery of 217 planets to date (May 2024).

The main advantage of this method is its capability to detect Earth-like planets orbiting Sun-like stars at distances between 1-10 AU. As a matter of fact, microlensing is the only proven method capable of detecting low-mass planets with wide orbits. This is something that traditional techniques such as the radial velocity and transit methods cannot do.[25]

Solar Gravitational Lens (SGL)

Detecting those low-mass planets is an impressive achievement, but what if I tell you that a variation of the microlensing method might potentially give us the first-ever image of the surface of an Earth-like planet around a Sun-like star? This is what Dr. Turyshev, a Research Scientist at Jet Propulsion Laboratory (JPL) at NASA and his team are proposing.[26] These researchers intend to use the Sun as a lens capable of focusing the light from a distant source. The concept is called *Solar Gravitational*

Lens (SGL) and takes advantage of the natural ability of the Sun to focus light from a dim and distant source. As we explored earlier, light gets distorted by a strong gravitational field like the one a star like the Sun produces, causing the trajectories of photons to be bent. In physics, this bending of light is called *refraction,* and according to Einstein's General Relativity, gravitation induces refractive properties in spacetime. The refraction of light phenomenon can be visualized by placing a pen in a glass of water. The pen seemed to have been broken, but what is happening is that when photons move from air to water, they deviate from their original path, making you believe that you need to buy a new pen.

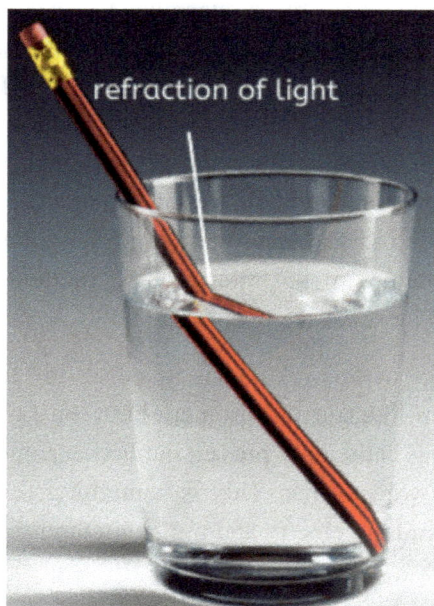

Fig. 4.24 Light is bent when moving from one medium (air) to another (water), causing the illusion that the pen is broken.

The key to the Solar Gravitational Lens method is to place a modest telescope of around 1 meter in the region where the bent photons converge; this place is called the *focal point.* Such an arrangement would allow astronomers to create a direct megapixel imaging of an exoplanet. This means that we could see clouds, continents, oceans, and even identify the different elements in the atmosphere. The physics is solid, but there is only one issue: the proponents have calculated that the focal point, or the place where the telescope needs to be placed, is at around 547 AU from the Sun.

With our current propulsion systems, getting to that point will take decades. For example, Voyager 1, launched in 1977, has traveled approximately 160 AU.

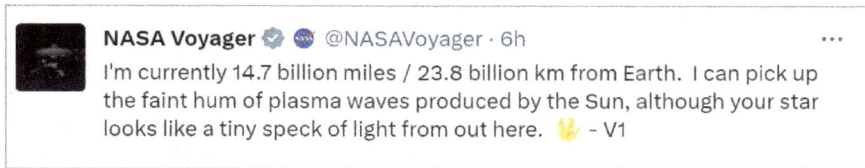

NASA Voyager ☑ 🛰 @NASAVoyager · 6h   ···
I'm currently 14.7 billion miles / 23.8 billion km from Earth. I can pick up the faint hum of plasma waves produced by the Sun, although your star looks like a tiny speck of light from out here. 👋 - V1

Fig. 4.25 A tweet posted by the official NASA Voyager X account on the 10th of May, 2023. Voyager 1 is the first spacecraft that entered interstellar space and the farthest man-made object.

This indicates that Voyager 1, traveling at a speed of 61,500 km/h, has taken 46 years to traverse a distance nearly one-fourth of what the proponents of the solar gravitational lens method suggest for the placement of their telescope.

Voyager 1's propulsion system is based on three radioisotope thermoelectric generators containing 24 pressed plutonium-238 oxide spheres. This propulsion system is used for trajectory correction maneuvers and adds little to the spacecraft's speed. Voyager 1's forward motion combines the original high-speed launch away from Earth, followed by a gravitational slingshot provided by Jupiter.

Traveling at the same speed as Voyager 1's would take 184 years to get to the proposed focal point. Not good enough for all of us who want to see an image of a beach from another Earth during our lifetimes.

Thankfully, Dr. Turyshev and his collaborators have other propulsion systems in mind. The researchers intend to use solar sailings as a propulsion mechanism.[27] Solar sailings work by taking advantage of the continuous push from the solar radiation. With a propulsion mechanism like this, a spacecraft could travel between 5 - 10 AU per year, which would allow it to reach the focal point in around 5 - 10 years; this is not science fiction. In 2010, the Japan Aerospace Exploration Agency (JAXA), demonstrated the viability of the solar sailing propulsion method during a mission to Venus. The mission, named Interplanetary Kite-craft Accelerated by Radiation Of the Sun (IKAROS),[28] became the first spacecraft to use solar sailing as the main propulsion mechanism.[29]

Fig. 4.26 A scale model of the IKAROS spacecraft. The solar sail is thin-film solar array designed to take advantage of the continuous solar radiation pressure. Credit: Czech Wikipedia user Packa.

The SGL methods fall under non-traditional ways to detect and image exoplanets. Such methods are not only the most exciting but also have some advantages over the most traditional ones. Let's keep exploring these non-traditional methods.

Pulsar Timing Variations

In 1992, Aleksander Wolszczan and Dale Frail changed the course of astronomy when they detected the first confirmed exoplanet using the pulsar timing variations technique. Pulsars, as we discussed in Chapter 1, are rapidly rotating neutron stars that emit intense electromagnetic radiation as they rotate. Astronomers detect these emissions as regular and precisely timed pulses. As we discussed previously, these pulses are so regular that a GPS-like system called X-ray pulsar-based navigation and timing (XNAV) has been proposed to help spacecraft in outer space orientate themselves. However, if a planet orbits a pulsar, the gravitational effect exerted by the planet causes slight yet regular variations in the pulsation period, which can be detected.

Pulsars are the remnants of a massive star. Before reaching this final stage, these stars undergo a series of explosions, collapsing events, and expansions. Such tumultuous events likely perturb the orbits of any potential planets around their parent pulsars, or in the worst-case scenario, the planets could be destroyed. Such conditions are why astronomers don't generally expect to find too many planets around pulsars. It may also explain why Wolszczan and Frail were not awarded the Nobel Prize in Physics alongside Mayor and Queloz in 2019.

Similar to the radial velocity method, pulsar timing variations favors detecting massive planets close to the host pulsar. This technique is even more sensitive to planets that orbit *millisecond pulsars*, which astronomers refer to as pulsars that rotate with periods of only a few milliseconds. The variations in the pulsar periods help astronomers determine the mass, distance from the pulsar, and orbital period of the planet.

Pulsation Timing Variations

I did not make a mistake; this method's name contains the word "pulsation", which is similar to the word "pulsar", from the previous method. Despite their similar names, the way these methods work is quite different. To understand how the Pulsation Timing Variations detection technique works, let's briefly divert our attention to a particular special type of star. Astronomers have known for a while that some stars vary their brightness on a regular basis. A star can vary its brightness for several reasons. For instance, stars can be part of a binary system, and their brightness changes due to the regular eclipses caused by their companions. A star's brightness can also vary due to surrounding material, such as dust and gas. However, a specific category of stars called pulsating stars exists. These pulsating stars dim and brighten as their surfaces expand and contract in a periodic fashion. This whole cycle is known as the *pulsation cycle* of the star, and the time between two consecutive peaks in brightness is the *pulsation period.* The most famous of such stars is Mira, discovered in 1595 by the Astronomer David Fabricius (1587-1615). Mira's luminosity changes by a factor of 100 over a period of 332 days. Further observations found that other stars also change their luminosities over long periods of time. Not surprisingly, such stars are known as *long-period variables* and exhibit pulsation periods between 100 and 700 days.

On the other hand, there are pulsating stars that change their luminosities over shorter periods of time. In 1784, John Goodricke (1764-1786), found that the star δ Cephei (δ is the lower-case Greek letter Delta) varies its brightness over 5 days. This star is the origin of the Cepheid stars category, which vary in brightness over 1 to 100 days.

What are these pulsating stars? These are intermediate mass dying stars, or stars that have consumed most of their original hydrogen supply. Just before transitioning to red giants, at extremely high temperatures, their still-burning hydrogen cores are surrounded by an envelope of helium (as discussed in Chapter 1).

Let's quickly remember that electrons and protons are particles in an atom's nucleus. Electrons have a negative charge, and protons have a positive charge, effectively causing atoms to have a neutral charge. In the case of the helium element, two electrons and two protons are in balance. Such a balance, however, can be disrupted. Inside a star, as temperatures rise, electrons are stripped off from the atoms of the helium atoms surrounding the core. This results in the atoms acquiring a net positive charge. Atoms with a positive charge are known as *ionized atoms* in physics and chemistry. Ionized helium atoms have the ability to prevent the transmission of light. The degree to which a material or substance impedes light transmission is referred to as its *opacity*. The higher the ionization level in a helium atom, the more opaque it becomes. If two electrons are removed from an atom, it is known as *double-ionization*, whereas the removal of a single electron is called *single-ionization*. Double-ionized atoms are, therefore, more opaque than single-ionized ones.

Remember that during the life of a star, gravity and pressure are in equilibrium. At later stages of a star's life cycle, this balance between gravity and pressure is not sustainable anymore. Gravity starts winning and compressing the star, increasing its temperature, and causing the helium atoms in the envelope to become double-ionized, reducing the star's brightness. As opacity increases, the temperature and pressure also increase. Given that the energy carried by light is absorbed, the outer layers of the stars expand, allowing the pressure to overcome gravity. As the star expands, the outer layers cool down, allowing the helium atoms to start capturing electrons again. However, temperature is still hot (after all, this is a star we are talking about), and the helium atoms can only capture a single electron. Therefore, they go from being double-ionized or not having any electrons at all, to be single-ionized. The less ionized the helium atoms become, the less opaque, or conversely, the more *transparent* they become. This increased transparency allows more light to penetrate through the star's outer layers, making the star brighter. Consequently, as the temperature drops even more, gravity takes over again. Then, gravity causes the star to compress, temperature increases, double-ionization occurs, the helium atoms become more opaque, and the whole pulsation cycle starts again. This pulsation mechanism is known as the *kappa opacity mechanism*, or simply, the *kappa mechanism*.

Okay, all this is cool, but why is it important? Well, it is important because of what the female astronomer Henrietta Swan Leavitt (1868-1921) discovered while working as a "human computer" for Edward Charles Pickering (1846-1919) at Harvard University. Her very boring human computing job involved comparing two photographs of portions of the sky at different times and identifying stars that have suffered variations in their brightness. Henrietta went above and beyond what she had been tasked to do, and after identifying around 2400 Cepheids, decided to study the nature of such stars. She noticed that the more luminous a star, the longer its pulsation period. Her research led to the discovery of a mathematical relationship between the actual luminosity of a Cepheid star and its pulsations periods, known as the *period-luminosity relation.* Something worth noticing is that there is a difference between the luminosity that astronomers can measure— also known as the brightness— and the actual luminosity—also known simply as luminosity— of a star. Astronomers refer to the former as the *apparent magnitude*, and to the latter as the *absolute magnitude.* Contrary to the brightness (the measured luminosity), the actual luminosity of an object does not depend on how far away it is or if there is stuff like gas or dust between the instrument and the object. The actual luminosity is the real luminosity, hence the name *absolute*. This is a game-changer because, from measuring the pulsation period of a variable star, scientists are able to determine its absolute magnitude. It is also important to note that the measured luminosity of an object decreases with the distance. Astronomers employ a tool called the *distance modulus,* which describes the relationship between the measured luminosity, the actual luminosity, and the distance at what the object is. However, the actual luminosity values for objects were unknown and elusive until Henrietta's period-luminosity discovery. By having a method to determine the real luminosity of an object, the distance at what that object is can be determined. Henrietta's discovery, also known as Leavitt's law, is one of the most important discoveries in Astronomy.

Prior to Henrietta's research, astronomers were limited at determining distances for objects using parallax angles they could resolve. For the technology at the time, this meant objects around a few hundred light-years from Earth. By using Henrietta's period-luminosity method, astronomers could suddenly calculate distances of objects up to 200,000 light-years from Earth. This literally expanded our view of the universe. Not a bad discovery for Henrietta who, being a woman, was not allowed to even touch a telescope at Harvard back then.

Enough of history already. The Pulsation Timing Variations method works by finding pulsating stars and small differences in their otherwise stable oscillation periods. If those small differences are periodic, they are mostly due to the presence of a secondary body with a low mass when compared to the pulsating star. This low-mass feature rules out another star, as in a binary star configuration. A low-mass object that can still noticeably disturb the star's orbit is more than likely to be a planet.

Only two planets have been discovered using this method at the moment of this writing. The discovered planets orbit stars that pulsate due to the previously discussed kappa mechanism. The first planet, with a minimum mass three times the mass of Jupiter, was reported[30] in 2007 orbiting the subdwarf B star V391 Pegasi, which is at a distance of about 4,000 light-years from Earth. There are two types of subdwarf stars: stars in the stage between the red giant and the white dwarf phases and stars that lie just below the main sequence and are cool main-sequence stars of very low metallicity.

V391 Pegasi belongs to the first type of subdwarfs. These subdwarfs are located on the *extreme horizontal branch* of the Hertzsprung-Russell (HR) diagram (see Chapter 1). Horizontal branch stars are those that are in the stage in the stellar evolution just after the red-giant phase. The term *extreme* refers to the fact these stars are the hottest stars in the horizontal branch, with temperatures around 25,000 Kelvin.

Subdwarf B stars have incredibly stable oscillation periods, making it easier to perceive small differences when a secondary body is present. V391 Pegasi b is the first planet detected orbiting a post-red-giant star, which may affect the longevity of a possible alien civilization (more of this in Chapters six and seven). Subdwarf B stars, a type of variable star, belong to the spectral class B.

The second planet discovered via the pulsation timing variations was reported in 2016.[31] The planet, which is 12 times as massive as Jupiter, orbits the main-sequence A-star KIC 7917485 at about 4,500 light-years from Earth. Like other pulsating stars such as the Cepheids, KIC 7917485 is located in the *instability strip* in the HR diagram. Stars within this instability strip are usually of spectral type A or F and exhibit variations in their luminosities due to the pulsations of their outer layers.

As we have seen, variations on an otherwise periodic event can help detect planets. Two other detection techniques that rely on this concept are the *Transit Timing Variations* and *Eclipse Timing Variations*. Let's explore these in more detail.

Transit Timing Variations (TTV)

In Chapter 3, we introduced the most successful detection technique: the transit method. With an impressive 75% of all the discovered planets, this method relies on observing a planet as it passes in front of its parent star, causing a dimming in the star's light. Effective observation of this phenomenon requires the observer to be positioned at an angle that allows them to detect such a dimming in the star's luminosity. Transits occur with remarkable consistency in systems where a single planet orbits a star, following an almost perfectly regular pattern. Transits consistently happen "on time" as the orbit of the planet is not affected by any other bodies.

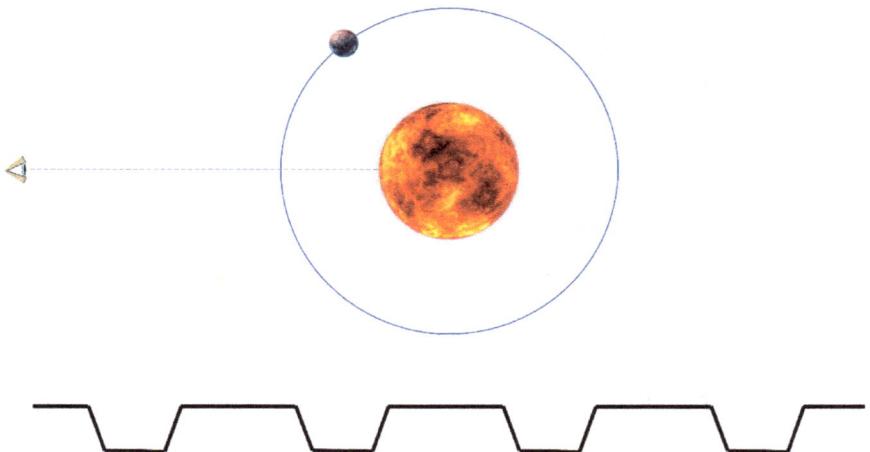

Fig. 4.27 A single planet orbiting a star displays a regular pattern for its transits, with those always occurring "on time". Credit: image by the author.

However, when multiple planets orbit a star, they interfere gravitationally, causing transits to sometimes happen early or late.

But why does this occur? Let's remember Kepler's second law, which states that a planet moves slower the farther away it is from its host star. Conversely, when a planet gets closer to the star it speeds up. For a given planet, another planet closer to the parent star is called an *inner planet*. On the other hand, a planet farther from the host star than a given planet is called an *outer planet*. For instance, Mercury is an inner planet for Earth because it is closer to the Sun. On the contrary, Mars is an outer planet as it is farther from the Sun.

Inner planets move faster than outer planets due to their closer proximity to the parent star, as they are more influenced by the star's gravitational field. However, the gravitational fields of the planets themselves result in either an acceleration or deceleration of other planets. If the inner planet's orbit is at a given position, behind the outer planet's orbit, the inner planet will gravitationally attract the outer one, slowing it down and causing a slight delay in the occurrence of the next transit of the outer planet. Consequently, the transit occurs later than expected. On the contrary, if the inner planet overtakes and positions itself in front of the outer planet, its gravitational influence will accelerate the outer planet, causing the next occurrence of the transit of the outer planet to be slightly earlier than expected.

For example, as shown in Figure 4.28, if planet A always transits its parent star every 60 mins, transit timing variations caused by an inner planet, planet B, will result in planet A's next transit occurrence being early by 10 mins, or late by 10 mins. I refer the reader to NASA's Transit Timing Variations animation[32], which can help further understand this technique.

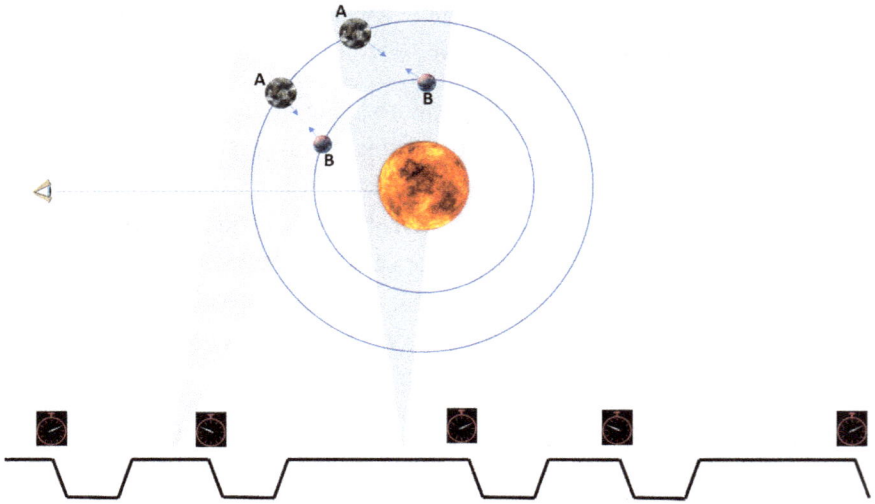

Fig. 4.28 When the inner planet accelerates the outer planet (B in front of A), it causes the outer planet's next transit to occur earlier than what it was expected. If the inner planet decelerates the outer planet (B behind A), the outer planet's next transit will occur later than anticipated. Credit: image by the author.

At present, 29 planets have been detected using the TTV technique. As this method relies on the gravitational effects caused by another planet, it allows the determination of the masses of the planets involved in the observation. Something that the normal transit method by itself is not capable of determining. The downside is that you have to observe multiple transits. This is not ideal for a couple of reasons. First, telescope time is quite precious, and pointing your instrument at a single target for a long time is usually impractical. Second, and consequently, this method is ideal for systems with planets that have short orbital periods, as such systems facilitate the observation of multiple transit occurrences within a short period of time.

A very interesting application of TTV is the potential discovery of exomoons, which are moons orbiting exoplanets. Over 70 candidates have been identified, but, as of May 2024, no exomoon have been confirmed. Exomoons can affect the orbital velocity of their host planet around the parent star. The presence of an exomoon will cause the planet's orbital velocity around the star to accelerate or deaccelerate, resulting in a longer or shorter transit duration. In other words, an exomoon will cause *transit duration variations*. By measuring such transit duration variations, the presence of an exomoon can be inferred.[33]

Exomoons are extremely interesting, not only because they can help astronomers better understand the formation of extrasolar planetary systems but also because it could help us explore the potential habitability of the cosmos. For instance, in the solar system, the best prospects for worlds capable of sustaining life are Europa and Enceladus, icy moons of Jupiter and Saturn, respectively. We will explore this topic further in Chapter 7. For now, let's continue with the Eclipse Timing Variation technique, which is very similar in principle to the TTV method.

Eclipse Timing Variations

This method aims to identify planets orbiting in binary star systems, where two stars revolve around a common center of mass. Although such systems may seem irrelevant to us because our planetary system contains only a single star, about 85% of all stellar systems in the universe are estimated to be binary or even exhibit multiple star configurations.

Fig. 4.29 The solar system only has one star. Apologies for the bad joke. I could not resist. Credit: image by the author.

In Chapter 1, while describing the star formation process, we explored the process of *fragmentation*. Essentially, star forming molecular clouds do not have a uniform density. There are fragments or smaller clouds within the larger cloud that are denser than others and experience a disequilibrium between the force of gravity and the thermal pressure. This disequilibrium results in gravitational instabilities of these smaller clouds, causing them to become self-gravitating cores and collapsing under their own weight. If the initial mass of these smaller clouds is large enough, they start fusing hydrogen into helium, hence becoming individual stars, thus leading to the formation of a binary or even a multi-stellar system.

As an anecdote, it is rare indeed that the Sun does not apparently have any siblings or at least not close-by ones.

The AMBRE project[34] is one of multiple efforts to find the Sun's brothers (or sisters). Potential solar siblings are stars that formed in the same cluster as the Sun and would have a similar chemical composition, as they would have originated from the same molecular cloud. The AMBRE project has collected a large spectra database of stars close to the solar system to determine if their chemistry matches that of the Sun. The team has found a candidate: the star HD 186302, with an estimated age of 4.5 billion years, the same as the Sun. HD 186302 is also a G-type main-sequence star located at about 185 light-years from Earth; it exhibits a similar chemical abundance, surface temperature, and luminosity. Further studies and analysis are required to declare that the Sun has a sibling somewhere out there.

When binary stars don't have any planets revolving around them, the two stars orbit around the system's center of mass without any external interference, eclipsing one another. This type of binary is known as *eclipsing binaries*. Astronomers can observe these eclipsing binaries if the orientation of the orbits of each star is aligned along our line of sight. In such a setup, astronomers can predict the system's orbital evolution with high precision. For stars with orbital planes oriented with our line of sight, one star will eclipse the other with extreme regularity, and we will observe that eclipses occur simultaneously and have the same durations. However, given that there are so many binary stars, it is not uncanny to think that planets could orbit those systems. Planets that revolve around two stars in a binary system are known as *circumbinary,* and interfere with the binary orbit, causing changes or variations in the period of the binary eclipses. Hence, the name Eclipse Timing Variations (ETV).

NASA lists 17 planets detected using this method. The first two planets, discovered in 2011, required the analysis of data spanning around 27 years.[35] These two planets orbit a system comprised of a white dwarf and a low-mass star. Essentially, planet hunters using the ETV method construct a model where they determine and predict the eclipses on a binary system. Any observed periodic deviation is used to infer the presence and parameters of an orbiting planet. For the discovered planets, astronomers detected two very distinctive periodic deviations. The first one with a duration of approximately 5.25 years, and the second one, with a duration of 16 years. These are indeed the times that each planet takes to complete a full orbit around the binary system or the planets' periods. The longer the analyzed data spans, the better the chances of finding unequivocal periodic deviations. However, a planet orbiting the binary system is not the only mechanism that can cause regular eclipse timing variations. Magnetic cycle mechanisms can also cause such periodic deviations. One of these magnetic cycles is referred to as the *Applegate mechanism* in honor of Douglas Applegate, the astrophysicist who proposed it in 1992.[36] In essence, this mechanism describes how one of the stars in the binary system can undergo a magnetic activity cycle that can potentiate a change in the shape of the star, becoming non-spherical. This change in shape can alter the gravitational interactions between the two stars in the system, altering their orbital period. Researchers rule out the Applegate mechanism as the mechanism causing the periodic deviations observed on a given binary system by considering the energy radiated by the stars. Astronomers can calculate with high confidence the amount of energy required to cause a period change. If the observed amounts don't correspond with their theoretical models, the Applegate mechanism is discarded. This methodology was followed by the reporters of the first two planets detected using ETV. However, more observations are required to dispel any doubts.

Orbital Brightness Modulations

We learned that astronomers can infer the presence of a planet due to the dimming of the light when it passes in front of its host star —the transit method. We also learned that for such transits to be detected, the planet, the host star, and the observer need to be along the same line of sight. However, if the planet-star system is face-on from the point of view of an observer, the observer won't be able to detect any variation in the brightness of the star (see Chapter 3 – Transit Method). Nevertheless, if the planet is massive enough and in a close-in orbit around its parent star, variations in the brightness of the star due to gravitational interactions between the two bodies can be detected. These variations in brightness, or orbital brightness modulations, is what astronomers employ to discover any potential non-transiting planet.

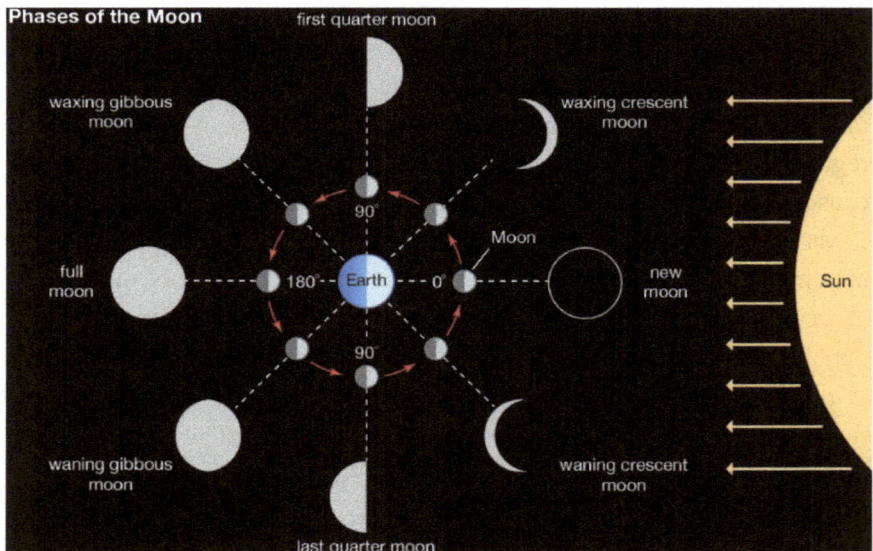

Fig. 4.30 The amount of light that the Moon reflects from the Sun varies as it revolves around the Earth. Credit: Britannica.

Similarly, a planet can influence the total brightness of a planet-star system based on the amount of light it reflects while orbiting its parent star. We are all familiar with this phenomenon when observing the Moon. As the Moon revolves around the Earth, it reflects varying amounts of sunlight, and its brightness varies during different lunar phases. During a Full Moon, it reflects the most light, while during the third and first quarters, it only reflects half of what it does at Full Moon. During the Waning and Waxing phases, approximately one quarter of the light is reflected, compared to the Full Moon.[37]

The likelihood of detecting these brightness variations in a planet-star system is higher if the planet is on a close-in orbit and possesses a high capacity for reflecting starlight or high albedo. We explored the albedo concept when discussing the Direct Imaging technique. If the observed brightness variations of the system are periodic, this can be a sign of the presence of a planet.

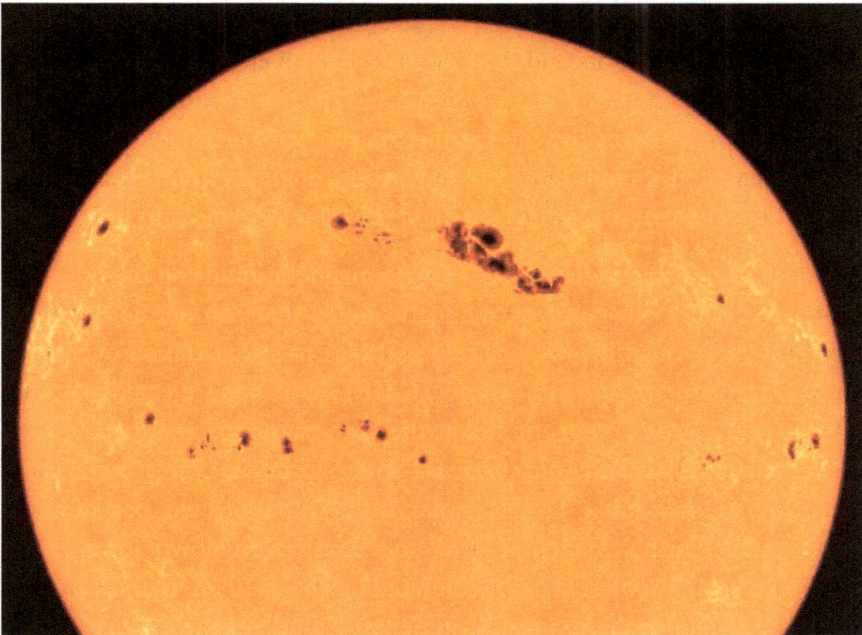

Fig. 4.31 Sunspots photographed by NASA's Solar and Heliospheric Observatory (SOHO). Sunspots look darker due to their lower temperature compared to the rest of the photosphere. Credit: SOHO/ESA/NASA.

However, other mechanisms can cause such periodic changes in brightness. For example, cyclic magnetic activity is common on stars. We are quite familiar with this phenomenon right here in the solar system. One of the most prominent features of the Sun is the well-known sunspots. Sunspots result from an 11-year magnetic cycle and appear as dark spots or patches in the Sun's visual surface, the *photosphere*. A sunspot is created when a strong magnetic field prevents energy from the interior of the Sun from being transported to the photosphere. When this happens, the temperature at that particular region is lower than the rest of the surface—that region is cooler, which results in an apparent darker spot. When we say that sunspots are cooler than the rest of the photosphere, it does not mean that we can go there and have a nice picnic day. The temperature of the darkest (coolest) region of a sunspot is typically between 2,700 to 4,200 Kelvin (approximately 2,400 to 4,000 degrees Celsius), which is still considerably cooler compared to the temperature of the photosphere, at around 5,500 Kelvin (5,300 degrees Celsius).

The number of sunspots and their sizes at any given moment varies with the intensity of the magnetic field that the Sun is experiencing. During the 11-year sunspot cycle, the number of sunspots increases at the peak and decreases at its minimum. Such periodic changes in the number and sizes of the sunspots cause variations in the total brightness of the Sun.

Sunspots are visible indicators of intense magnetic fields on the Sun. The spots are filled with complex magnetic fields that can act as catapults, throwing large quantities of charged particles to space. These events are known as *Coronal Mass Ejections* (CME). When the ejected particles interact with Earth's magnetic fields, they can trigger *geomagnetic storms* (*Geostorm*), also called *solar storms*. Solar storms are most common at the peak of the Sun's 11-year solar cycle. The ejected particles are captured by Earth's magnetic field and accelerated down towards the north and south poles. The accelerated particles then collide against atoms in the atmosphere, transferring energy and causing them to *excite*. *Excitation* is a process whereby the electrons in an atom gain energy and transition from a lower energy level to one of higher energy. Electrons then release this recently gained energy as photons (light), causing the magnificent auroras in both hemispheres. These auroras are referred to as *aurora borealis* when observed in the northern hemisphere and *aurora australis* when occurring in the southern hemisphere.[38]

Fig. 4.32 An impressive aurora australis is captured from Eaglehawk Neck in Tasmania, Australia, in May 2024. Credit: Photo by Sean O' Riordan.

The most intense and disruptive solar storm ever recorded is known as the "Carrington Event" which peaked in September 1-2 in 1859.[39] The solar storm was so intense that it caused auroras to be visible as far south as Mexico and Hawaii and disrupted telegraph systems throughout Europe and North America. As our current society relies so heavily on satellites, power grids, and cellular phone networks, a new occurrence of such events could profoundly impact our daily lives. For such a reason, astronomers and researchers continue to monitor these events with the hope of being able to predict with more certainty when the next big geostorm will hit Earth so we can prepare accordingly.[40]

Magnetic cycles on stars exhibit relative periodicity, but sunspots-like features are not permanent. The number and sizes of sunspots can vary considerably, leading to variable effects on the overall changes in the brightness of the Sun. Changes in brightness caused by a planet tend to be more periodic and stable. This distinction provides researchers a way to tell them apart.

NASA lists only nine planets detected via the Orbital Brightness Modulations (OBM) method. Scientists usually collect large sets of light curves collected with instruments like Kepler and then analyze them using a set of algorithms. The planet Kepler-76 b[41] was discovered using the OBM method and then confirmed with the radial velocity technique.

A number of characteristics can be determined from the effects of a planet on the brightness of its host star. For instance, as the planet goes around the star, it moves towards or away from the observer. While discussing the radial velocity method, we already explored the effect that this movement causes in the stellar spectrum as perceived by an observer. The light received from a star being orbited by a planet will be periodically red-shifted or blue-shifted and this can help astronomers determine the mass of the planet. In addition, the gravitational interaction between a planet and its parent star can cause the shape of the star to deform into an ellipse, something astronomers call the *ellipsoidal effect*. Such a change in shape causes a variation in the measured brightness of the star. The ellipsoidal effect helps to determine the size of the planet and the mass ratio between the planet and the star. Other measurements, such as the reflected and emitted light from the planet, help scientists to understand the reflectivity of the planet's atmosphere, which can reveal the presence or absence of certain chemical compounds.

Disk Kinematics

Kinematics is a branch of physics that studies the motion of objects without considering the forces involved or the causes of the motion itself. The idea is to quantify the motion of an object by analyzing its change in position, velocity, and acceleration over time.

Applied to the exoplanets field, the Disk kinematics technique examines the motions of the gas and dust in the circumstellar disk surrounding a young star. By analyzing the kinematics of the disk, astronomers attempt to identify any possible motion patterns or signatures that may indicate the presence of a planet. For instance, as we explored in Chapter 2, gaps are a common feature in protoplanetary disks; the proponents of disk kinematics highlight that some gaps are the result of a forming planet interacting with the disk.

Only one planet has been discovered using this method.[42] The planet HD 97048 b which was detected by analyzing its effects in the gas surrounding the parent star. Researchers used the Atacama Large Millimeter/submillimeter Array (ALMA) observatory, an array comprised of 66 antennas. ALMA uses the interferometry technique, combining all the radio waves captured individually by each antenna, producing the same result as if they were a single 16-kilometer diameter giant telescope. ALMA observes in the portion of the spectrum between the far infrared and radio. The wavelength of these electromagnetic waves is on the order of millimeters or even sub-millimeters, which means that they have larger frequencies and, therefore, are considered low-energy waves. In protoplanetary disks, we find dense regions with high concentrations of dust grains and gas molecules. These dust particles are heated by the young star and emit thermal radiation at the millimeter and submillimeter wavelengths. The advantage is that light at these wavelengths does not interact with other particles or surfaces, as in, it does not scatter that much, nor does it get absorbed by gas molecules. This allows astronomers to have a peek within these dense regions.

In the case of HD 97048 b, astronomers inferred the presence of the planet by analyzing the gap and spirals in the gas surrounding the star. To calculate the planet's size, the researchers used simulations that varied the planet's mass. In these simulations, they also varied the number of orbits that the planet makes around the host star. In this case, they used 800 orbits, which is the equivalent of one million years. Then, they compared the results of the simulations to what they were observing and determined the mass of the planet to be around two to three times the mass of Jupiter. The size and the width of the observed and simulated gap gave researchers clues about the mass of the planet. A more massive planet will create a wider and deeper gap in the protoplanetary disk compared to a lower-mass planet. By varying different parameters in their simulations, such as the disk's viscosity (how thick a fluid is), and physical properties of the disk material, astronomers can also determine the planet's composition.

In these last two chapters, we have seen examples of the extreme curiosity, genius, and practicality of scientists' minds. All the discussed techniques are a testament to that persistence and inventiveness. We went from knowing only eight planets (the ones in the solar system) to thousands of planets, all discovered in the last couple of decades, and the planet count continues growing at a rapid pace.

With such numbers of planets, astronomers need to somehow classify those to help them understand them better. Classifying things is part of our human nature, and exoplanets are no exception. In the next chapter, we will explore the various categories astronomers have devised to make sense of their observations. This categorization aids in understanding the formation, configuration, and evolution of planetary systems, a branch of exoplanet study known as *planetary system architecture*.

Chapter 5
Exoplanets Classification

"She's a main sequence star, a lot like our own. Five planets…...and one of them: square in the habitable zone. A prime candidate."

— Ricks, Alien: Covenant (2017)

TL;DR

We use classification as a tool to better understand the nature of things. South African Bishop Desmond Tutu (1931-2021) Nobel Peace prize 1984 wisely stated "there is only one way to eat an elephant: one bite at a time". Often, it is more convenient to build up our understanding using a bottom-up strategy, considering small pieces rather than approaching complex concepts and structures from a macroscopic view. Considering this, we construct categories instrumental in grouping similar entities to understand them better. For instance, take the complex concept of planetary system architecture in a stellar system. We can improve our understanding of this by breaking down large populations of planets into small "bites" or categories.

Planets within the solar system can be popped into two very large buckets: small rocky terrestrial worlds (Mercury, Venus, Earth, and Mars) and worlds with humongous gaseous atmospheres (Jupiter, Saturn, Uranus, and Neptune). Unsurprisingly, astronomers continue with a similar categorization technique for discovered worlds beyond the solar system. However, when trying to do so with the discovered planets, astronomers realized that not all stellar systems out there are like ours. Exoplanets come in so many "flavors" that new categories had to be established. The most practical way to classify these celestial objects is based on the exoplanet's sizes and masses. Hence, this is the most widely adopted classification technique. Planets are classified into four different categories: i) **Gas Giants**. Constitute 30% of all discovered exoplanets to date. Their masses vary from half to 13 times the mass of Jupiter, and their sizes vary between one and 1.7-times Jupiter's size. The most famous of such planets is the first ever exoplanet discovered orbiting a main sequence star: 51 Pegasi b; a gas giant with a mass 0.46 times that of Jupiter and 1.27 times its radius. Gas giants, as their name implies, are predominantly made up of gas, specifically, a very thick atmosphere of helium and/or hydrogen surrounding a rocky or molten core. ii) **Neptunian or Ice Giants.** In the solar system, the large distances that Neptune and Uranus are located from the Sun cause gases to solidify as ice. At these low temperatures, gases behave like solid rocks and therefore, these planets are known as *Ice Giants*. With sizes ranging between two and a half and up to four times the size of Earth, ice giants are mostly composed of water, ammonia, methane, and carbon dioxide. However, exoplanets classified as Neptunians are done so due to their similar sizes and masses to Neptune. In other planetary systems, Neptunians are not located at the large distances from their host stars that we observe for Uranus and Neptune. Therefore, their composition seems to depend on their proximity to their host star. These planets make up 35% of all the detected exoplanets. iii) **Terrestrial.** With Sizes between half and twice that of Earth and masses equaling Earth's, these planets are made of a solid rocky core and a thin or, in most cases, inexistent atmosphere. Due to their small sizes and difficulty finding them, less than 4% of all detected planets up to date belong to this category. These planets represent our best chance of discovering life elsewhere, at least in the forms we are familiar with. An atmosphere, potentially due to comet impacts or early volcanic activity, could create conditions that facilitate the development of vegetation and complex life forms. This is the reason many astrobiologists are focusing their efforts on the TRAPPIST-1 planetary system and its seven rocky planets. For some of the planets in the TRAPPIST-1 system, due to their distances to their host star, there is a possibility of liquid water flowing in their surfaces, implying the potential to support life. iv) *Super-Earths*. These are planets of anything between two and two and a half times the size of Earth, and up to 10 times Earth's mass. A kind of misleading name if you ask me. Nearly 31% of all detected planets belong to this category. Super-Earths can be rocky, but they can also be made of gas, or a combination of gas and rock. When a Super-Earth is found to be composed mostly of gas, astronomers call those *mini-Neptunes* or *sub-Neptunes*, as they are smaller than the planet Neptune in the solar system but have similar compositions.

Researchers also classify complete planetary systems into four categories: i) *Similar*. Where all the planets have masses that are approximately similar to each other, ii) *Mixed.* Planetary systems where it does not exist a discernible regular pattern in the mass distribution of the planets regarding their distances to the parent star, iii) *Anti-ordered*. Where the mass of the planets decreases according to the distance from the star, and iv) *Ordered.* Planetary systems where the mass of the planets increases according to their distance from the star.

Most of the time, a planet's size or mass is not enough to describe it properly. An example of this is the presence of gas planets very close to their host star. These planets are called *Hot Jupiters*, as their surface temperatures are extremely high due to the proximity to their parent stars; Hot Jupiters exhibit sizes similar to or larger than Jupiter. These planets are nothing like what we have in the solar system. They are so close that a full revolution around their host star does not take them years but days or even hours. This difference in their periods is used by astronomers to categorize them. Very-Hot Jupiters (VHJ), for instance, are planets that have periods between two and three days. Ultra-Hot Jupiters (UHJ) complete a whole revolution in a matter of hours. The record goes to the WASP-19 b planet, which orbits its parent star in only 18.9 hours!

Astronomers also have categories based on how many stars an exoplanet orbits, as planets can orbit more than one. A *circumbinary planet* orbits a binary system, a *circumtrinary planet*, orbits a system with three stars, and a *multi-stellar planet* belongs to a system with more than three stellar objects. On the other hand, you have orphan planets that don't have a parent star; the so-called *Rogue planets*. Rogue planets are detected using the gravitational microlensing technique or via direct imaging in the infrared. These planets are more than likely to be the result of gravitational forces or collisions in the early stages of planet formation of their former planetary system. Such collisions or gravitational influence of other planets resulted in these planets escaping from the gravitational field of the host star. Like errand spirits on a classic horror story, they are now free-floating through space, waiting for someone to detect them.

Finally, all the planets discovered so far are in close vicinity. That is, they are all located in the galaxy we inhabit, the Milky Way. But nothing indicates that planet formation is only restricted to our galaxy. Therefore, astronomers expect planets to be a common feature of stellar systems in the universe. However, given the vastness of the universe, detecting planets that inhabit other galaxies is extremely cumbersome. These extragalactic planets are still elusive and difficult to detect. However, objects like the one orbiting the X-ray binary M51-ULS-1 in the Whirlpool Galaxy (M51), located about 23.16 million light-years away, are promising candidates for such a category of planets.

Fig. 5.1 The different categories of exoplanets discovered so far. There is a clear bias in favor of big planets. Credit: NASA/JPL-Caltech.

Planetary System Architecture

Classifying things helps humans understand them. Creating categories according to certain characteristics hints at the nature and even origin of things. Categorization also contributes to recognizing patterns and provides a way to see how things relate or don't relate to each other. With the rise of large data sets, we can categorize systems in ways we've never thought possible. For instance, when COVID hit, scientists were able to classify how the virus affected the population based on their age, sex, weight, etc. This helped researchers understand the virus and provided them with ideas on how to come up with an eventual vaccine.

The field of exoplanets is not different. Now that we have more than 5,000 detected exoplanets, these are classified according to several characteristics. The goal is for this classification to shed some light on their forming process and fate and even help astronomers to determine how planets within the solar system came to be.

The field of *Planetary System Architecture* deals with studying the arrangement and characteristics of planets, asteroids, moons, and other celestial bodies in a stellar system concerning the central star. Up to 1992, scientists only had one sample of how stellar systems were formed and arranged. As more and more exoplanets are discovered, astronomers keep increasing their understanding of planet formation, which helps formulate new hypotheses. To this aim, planets are classified according to their composition, size, and proximity to their host star.

Composition

As we discussed in Chapter 2, the final composition of a planet depends on where it forms with respect to the parent star. Planets that form close to their host star tend to be rocky as all the gas surrounding their cores gets evaporated by the higher temperatures it experiences. In the solar system, examples of rocky planets include Mercury, Venus, Earth, and Mars, mostly made of metal and rock. Types of rock can include silicate rocks, which are composed primarily of silicate materials—combination of silicon, oxygen, and or more metals, and iron-nickel rocks. On the other hand, if a planet forms beyond the ice or snow line, temperatures are extremely low, and water and other molecules/atoms can remain in a frozen state. Planets at these locations contain large ice particles that behave like rocks. This extra frozen material makes those planets more massive than their rocky siblings and allows them to keep the accumulated gas.

It is, of course, not as easy as just simply saying that a planet is rocky or gaseous. Such a classification is mostly based on the density of the observed planet. Density is essentially a measure of how much mass is packed in a certain volume. As planets, and indeed exoplanets, are spherical in nature, astronomers use a sphere's volume to determine a planet's density. The volume of a planet can be calculated easily by knowing its radius and its mass. To this end, astronomers combine multiple detection techniques to determine the density of the planets. The radial velocity method indicates the mass, while the transit method helps in the determination of a planet's radius.

Rocky Planets

Rocky planets, also known as terrestrial or telluric, are mostly composed of silicon, magnesium, iron, carbon, and oxygen. Or in other words, these are planets which size is dominated by solid material. One way to subcategorize rocky planets is to distinguish between the ones that can retain an atmosphere and the ones that have thin or negligible atmospheres. The capacity for a planet to retain an atmosphere depends on the mass and the radius of the planet, as this has a direct impact on the escape velocity of the planet. We will explore the classification of planets based on their sizes later.

Gaseous Planets

Gaseous planets are large or super large compared to rocky planets. These planets possess huge amounts of hydrogen and helium, captured from the original nebular gas during their formation. Contrary to rocky planets, these planets' size is dominated by liquid or gas. Due to the amount of gas compared to their solid core, some of these planets have extremely low densities, even lower than water. We have an example of those in the solar system, the gas giant Saturn. Saturn's density is lower than water, and if it were possible to have an unthinkably large hot tub in which we could place Saturn, it would float. This is, of course, a quite unlikely experiment we can perform. Not only because I don't think we would be able to find any removalists to help us with such an endeavor, but because the gas surrounding Saturn will dissolve when in contact with water, leaving Saturn's solid core exposed. Saturn's core, which is rocky, is denser than water, and therefore, it will sink to the bottom of our imaginary reservoir of water.

Classification according to Size and Mass

I keep repeating this, but less than forty years ago, the only planetary system astronomers knew about was the solar system. Based on that sample size of one, astronomers inferred a lot of rules around planetary system architectures. For example, rocky planets are usually small and are closer to their host star. Gaseous planets are larger than rocky planets and are far away from the central star. Based on that understanding, it is not too hard to imagine why, in 1989, David Latham and his collaborators at the Harvard-Smithsonian Center for Astrophysics, using the radial velocity method,[1] dismissed the discovery of what they believed to be a giant planet orbiting at a very close proximity of its host star HD 114762. Unlike any of the planets in the solar system, this hypothetical planet, HD 114762 b, was 11 times more massive than Jupiter and orbited its host star at less than a tenth of the size of Jupiter's orbit, or 0.36 AU. Such a distance is similar to the size of the orbit of Mercury around the Sun; this was nonsense. In the solar system, giant gaseous planets have large orbits, and only the small rocky planets orbit close to the central star. Based on this information, they decided to report the discovery of a *brown dwarf* and only mentioned the possibility of this object being a planet in a speculative way. Fast forward 34 years, and we find that large-mass gas planets in close-in orbits are no longer a rarity. Astronomers have found thousands of those, and nobody currently doubts that such a planetary system configuration exists. Interestingly enough, new research has provided evidence about HD 114762 b being indeed a brown dwarf.[2] Its mass has been estimated to be between 82 and 139 times the mass of Jupiter, which is much larger than the masses found in the planetary domain. Regardless of the status of HD 114762 b, the moral of the story is that in science, we can't assume that we know all the answers; that is a good thing, as this is how science works and progresses. On this occasion, scientists learned their lesson, and in 1995, Mayor and Queloz decided to report the discovery of 51 Pegasi b , a planet with a period of only 4.2 days and a mass of half of Jupiter mass, equivalent to 150 the mass of Earth. This planet orbits its host star 51 Pegasi at 0.05 AU! We don't have anything like 51 Pegasi b in the solar system. From there, thousands of planets, like anything we encounter in our planetary system, have been discovered.

Suddenly, astronomers realized that new categories needed to be created. Let's explore these next.

Fig. 5.2 Artistic expression of 51 Pegasi b. Credit: NASA.

Gas Giants

The term "Gas Giant" was coined by science fiction author James Blish in his story *Solar Plexus* in 1952:

"A quick glance over the boards revealed that there was a magnetic field of some strength nearby, one that didn't belong to the invisible *gas giant* revolving half a million miles away."

A remarkable example of these types of planets is KELT-9 b, reported in 2017 and discovered using the transit method.[3] KELT-9 b is a planet that is hotter than most stars. It orbits the star KELT-9, which is a very young star, only 300 million years old and located 667 light-years from Earth. With a dayside temperature of 4,600 Kelvin (4,326.85 degrees Celsius), a mass nearly three times the mass of Jupiter and nearly twice the size Jupiter's size, this planet orbits its parent star with an orbital period of 1.5 days at a distance of 0.03 AU. Due to its proximity to the host star, the planet's atmosphere is constantly bombarded with high levels of ultraviolet radiation, possibly causing it to exhibit a comet-like tail of evaporated planetary material.

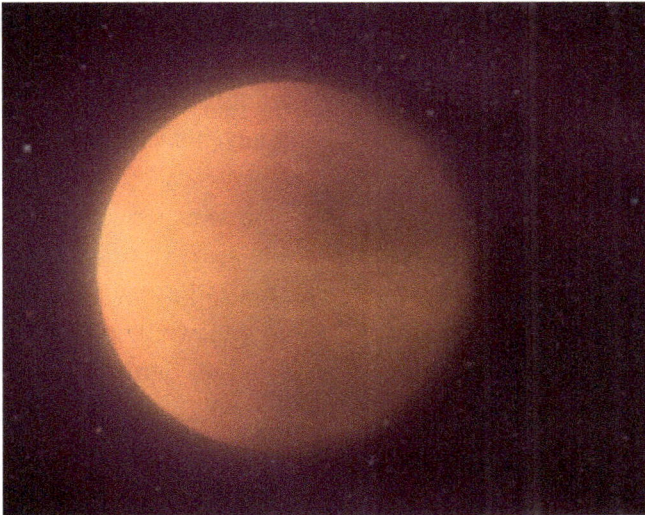

Fig. 5.3 Artist's concept of the gas giant KELT-9 b. One "year" on this planet is less than two days long. Credit: NASA / JPL-Caltech / Robert Hurt.

A Story of More Than 5000 Worlds

Gas giants typically have a rocky or molten core surrounded by a thick atmosphere mostly composed of helium and or hydrogen. In the solar system, Jupiter and Saturn belong to this category. The strange thing about this composition is that you cannot set foot on these planets as they don't have a surface. Saturn's core, for instance, has been estimated to extend to approximately 60% of the total planet's radius and only constitutes around 18% of the total mass.[4] In the case of Jupiter, the difference is even larger. It has been estimated that Jupiter's core's mass only accounts for up to 0.2% of the planet's total mass, whereas the planet core's radius only accounts for up to 0.07% of the total radius.

The close proximity of gas giant exoplanets to their central star, along with their enormous masses and radii, causes gravitational and brightness effects on the star, facilitating their discovery as discussed in Chapter 3 due to the nature of the radial velocity and transit methods. Therefore, it is not surprising that they account for 30% percent of all the exoplanets detected so far.

Typically, gas giants have sizes ranging from one to 1.7 times that of Jupiter, and their masses fall between half and 13 times that of Jupiter. The current understanding is that their enormous masses result directly from having formed far away from the star's gravitational influence and heat. This allows them to accrete gas around their large cores. Astronomers believe that if gas giants are formed around Sun-like stars, they do it during the first 10 to 100 million years,[5] which is considered a "short-term" process in the formation of planetary systems. The extreme pressures that the atmospheres of these planets experience cause amazing weather phenomena. For instance, the famous *Great Red Spot* in Jupiter is a large storm that has lasted for at least 193 years.[6] The clouds in this storm spin in an anticlockwise direction with winds that are more than twice as intense as a Category 5 cyclone on Earth. Using data from Hubble, we have recently learned that the winds are blowing more than 640 kilometers per hour.

Fig. 5.4 Jupiter's Great Red Spot is so large that three Earths could fit inside it. Credit: NASA, ESA, A. Simon (Goddard Space Flight Center), and M. H. Wong (University of California, Berkeley), and the OPAL team.

Another amazing weather feature for a gas giant observed in the solar system is the incredible hexagon-shaped storm in Saturn. The hexagon, which is wider than two Earths, is hypothesized to result from the interaction between Saturn's atmosphere and its fast rotation. Scientists have proposed that the hexagon-shaped storm is the result of a system of vortices, which are rotating columns of air that can form in fluids. In this case, the vortices are thought to be induced by Saturn's magnetic field. The hexagonal pattern emerges as a consequence of the interaction between the vortices and the winds generated by Saturn's rapid rotation.

Fig. 5.5 Saturn's hexagon-shaped storm located around the north polar regions as captured by the Cassini spacecraft in April 2014. Credit: NASA.

It is only expected that astronomers will be able to find weather features like those exhibited by the gas giants in the solar system or even more extraordinary phenomena in exoplanets of similar nature.

Neptunian Planets

In the solar system, Uranus and Neptune are located at 19 AU and 30 AU from the Sun, respectively, which is farther away than Jupiter and Saturn, located at 5 AU and 9.5 AU, respectively. For Uranus and Neptune, the larger distances at which they are from the Sun, have allowed water, ammonia, methane, and carbon dioxide to exist in a solid form known as *ices*. These ices behave like solid rocks, enriching the core and increasing the mass of Uranus and Neptune, allowing them to accumulate their surrounding gas. This is the reason these planets are known as *Ice Giants*.

The composition of those ices is what causes the beautiful blue that we see in the pictures of Neptune and Uranus. The atmospheres of these ice giants contain high concentrations of methane, which absorbs red light and reflects blue light. However, we can see that Uranus is paler than Neptune. Until recently, this was a mystery, given the similarities in the composition of their atmospheres. Researchers from the University of Oxford have found[7] that the paler color of Uranus is due to an extended layer of haze or light vapor in its atmosphere composed mostly of hydrocarbons and other organic compounds. This haze dilutes the blue color of Uranus and makes it look paler than Neptune.

Fig. 5.6 Uranus (left) and Neptune (right) are similar in size, mass, composition, and structures, but Uranus looks less blue than Neptune. Image taken by Voyager 2. Credit: (NASA/JPL-Caltech; NASA).

Exoplanets Classification

The current location of the ice giants poses a problem for our understanding of planet formation.

As we discussed in Chapter 2, the core accretion model is the most accepted hypothesis nowadays. In this hypothesis, terrestrial planets and the cores of gas giants grow by impacts or collisions of dust grains, which eventually form small rocks. These small rocks also collide, forming larger rocks known as planetesimals. This process continues until a solid core is formed. Gas giants' cores are large enough to gravitationally attract and maintain the gas in the original nebula. Small rocky planets do not have enough mass to keep the gases that surround them, and they end up with a very thin or non-existent atmosphere. In its purest form, the core accretion model primarily explains the formation of gas giants or smaller rocky planets based on the distance from the parent star. The model does not fully account for the formation of ice giants at their current location in the solar system. However, the model can be refined to include concepts such as core composition and planetary migration to help explain their existence. The root of the current location problem is that the collision accumulation process in the outer solar nebula is slower than in the regions where the gas giants formed due to the longer orbital periods and a reduced concentration of solid particles. Moreover, at 20 AU or 30 AU from the Sun, which is where these ice giant planets are in the solar system, the velocity that an object needs to escape the solar system—solar system escape velocity— is around 8 kilometers per second. According to simulations, this velocity is comparable to the velocities experienced by growing planetary embryos. At these speeds when embryos suffer the effects of mutual encounters, it results in those embryos either being ejected from the solar system or put into very eccentric cometary orbits. For these reasons, it does not look like Uranus and Neptune formed where they are currently located. The leading hypothesis that explains why they are where they are today indicates that they were formed originally between Jupiter and Saturn at the same time that these gas giants formed. Neptune and Uranus were then pushed outwards to their current location due to gravitational interactions with the larger planets.[8]

In the exoplanets realm, contrary to what we observe in the solar system with Uranus and Neptune, the majority of Neptunian-like exoplanets— planets with a similar size to Neptune— that astronomers have found so far are close to their host stars. However, like gas giants, Neptunian planets also have atmospheres comprised mainly of hydrogen and helium. Ice giants are planets with masses ranging from about 10 to 20 Earth masses and radii approximately between 2.5 to 4 times that of Earth's.

Fig. 5.7 Artistic impression of Kepler-1655 b. This Neptunian planet orbits very close to its host star, which is nothing like what we have in the solar system. Credit: NASA.

The large sizes of these planets and the proximity to their parent stars also make them "easier" to detect with our current techniques. It is unsurprising that nearly 35% of all exoplanets detected are categorized as Neptunians. For instance, the planet Kepler-1655 b, discovered in 2018,[9] is a Neptune-like planet that completes a full orbit around an F-type star located at 694 light-years from Earth in just 11.9 days. With the equivalent of nearly six times the mass of Earth, it is located at only 0.1 AU from its host star. Its size is more than twice the Earth's, and it was detected using the transit method. Being so close to the star, it is more than likely that its atmosphere could expand and evaporate.

Terrestrial Planets

NASA's definition of a terrestrial planet is one that is between half the size of Earth and twice its radius, and it has up to one Earth mass. In the solar system, rocky planets are Mercury, Venus, Earth, and Mars. Contrary to the gas and ice giants, these planets do have a solid surface. Therefore, they are good candidates in the search for extraterrestrial life.

As we have discussed in previous chapters, their smaller sizes make terrestrial planets difficult to find with the current detection techniques. Less than 4% of all the exoplanets detected to date are rocky planets. However, estimations from statistical studies[10] have indicated that only across our galaxy could be more than 10 billion terrestrial planets waiting to be detected and analyzed.[11]

Fig. 5.8 The solar system terrestrial planets. From left to right, Mercury, Venus, Earth, and Mars. Venus is almost the same size as Earth, but Mercury and Mars are smaller. Credit: NASA.

Rocky or terrestrial planets are mostly composed of rock, silicate, water, and or carbon. They can have atmospheres, continents, or even water if they are at a distance from their parent star where water can exist in liquid form on their surfaces.

Like Earth, most rocky planets have a core mostly composed of rock or iron and an atmosphere. They can have a solid or liquid surface depending on the distance to their host star and the different temperatures they experience, and the amount of radiation they receive. Contrary to gas or ice giants, the atmosphere of rocky planets did not originate from the original solar nebula. Their atmospheres are the result of comet impacts or what scientists call as *outgassing*. Outgassing refers to the release of gases from the interior of the planet through volcanic activity. Gases like water vapor, carbon dioxide, sulfur dioxide, and nitrogen oxides are expelled to the atmosphere during volcanic eruptions. Such components enrich a planet's atmosphere and can play an important role in the evolution of a rocky planet's atmosphere.

The solar system is comprised of a mix of rocky and gaseous planets. However, there is a planetary system out there where 100% of all the planets are rocky. TRAPPIST-1 is a planetary system that has seven rocky planets named TRAPPIST-1 b to TRAPPIST-1 h. These are rocky worlds with the potential that at least three can harbor liquid water on their surfaces.

Fig. 5.9 Artistic impression of the TRAPPIST-1 system. All planets are Earth-size planets with three of them in the star's habitable zone. Credit: NASA-JPL/Caltech.

The discovery of the TRAPPIST-1 system was announced by NASA in 2016.[12] The detection was made using NASA's Spitzer Space Telescope, which, like the JWST, sees the universe in the infrared part of the electro-magnetic spectrum. However, it does so at different wavelengths than those observed by the JWST. Spitzer's main science goal is to investigate objects closer to Earth such as dust clouds, protoplanetary disks, and exoplanets. JWST, on the other hand, is capable of covering Spitzer science goals and is better equipped to delve into the early universe by analyzing very young stars, galaxies, and planetary systems.

The TRAPPIST-1 system is extremely interesting, given the repercussions it has in the search for life. The system is located at about 40 light-years from Earth. If somebody sends a message to one of those planets, it will take forty years to reach the system and another forty years to receive any potential reply. That is, of course, if there is a "they", and if "they" decide to reply.

Since its detection, the TRAPPIST-1 planetary system has been breaking some records. First, at its discovery, it was the extrasolar system with the largest number of planets around a parent star with seven. As of today, the title now belongs to the Kepler-90 system with eight planets. However, TRAPPIST-1 is still the system with the largest number of planets in the habitable zone, with three. In addition, all the seven planets are Earth-size planets. This means that the three planets in the habitable zone can have an environment similar to the one we have on Earth, with large oceans and continents.

All the seven planets in the TRAPPIST-1 system orbit very closely to the host star with periods between 1.5 days for the closest planet, located at only 1.66 million kilometers from the star, and 20 days for the most distant, which is located at nearly 9 million kilometers. As a comparison, Mercury, the most inner planet of the solar system, orbits the Sun at nearly 58 million kilometers. This means that all the TRAPPIST-1 planets are on an orbit closer to the host star than Mercury, the closest planet to the Sun as illustrated in Figure 5.10.

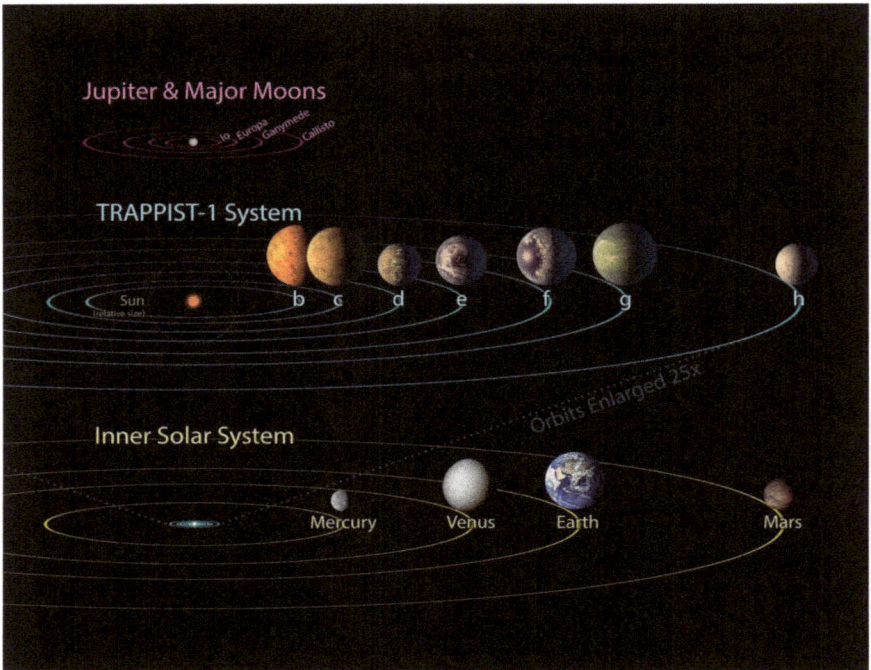

Fig. 5.10 The TRAPPIST-1 planetary system located at 40 light-years from Earth. All seven planets could fit inside the orbit of Mercury. Credit: NASA/JPL-Caltech.

Scientists believe there are three planets where water could exist on their surfaces. These planets, TRAPPIST-1 e, TRAPPIST-1 f, and TRAPPIST-1 g, respectively at 4.19, 5.53, and 6.73 million kilometers from their home star. You may be wondering: if they are so close to the star, how is it possible that water can potentially exist on their surface? Would not such a proximity to the star cause the water to evaporate? The thing is that the star TRAPPIST-1 is not the same type of star as the Sun. TRAPPIST-1 is an ultra-cool dwarf star with effective temperatures below 2,700 Kelvin (2,426.85 degrees Celsius). This relatively low temperature would allow planets to be close to the star and still have water in a liquid form on their surfaces.

Another peculiarity of the TRAPPIST-1 system is the fact that all of its planets are *tidally locked* to their host star. What this means is that all the planets always show the same face or hemisphere to the star. This is not something we are unfamiliar with. Given how close our Moon is to Earth, it is tidally locked to the Earth, so we always see the same face of the Moon at all times. This is what causes some people to call the side of the Moon we don't see as the "dark side", which is wrong. The Moon does receive light from the Sun on that side, so it is not dark. It is just that we can't see that side of Earth. Therefore, a better term for that side of the Moon, as seen from Earth, would be the "far side" or the "hidden side". This tidally locked condition happens for an object when its rotation period matches its orbital period around the parent body. This is typically caused by the proximity of the body to the parent body and how the rotation of the orbiting body is slowed down over time. In our case, the orbital period of the Moon, 27.3 days, matches the Moon's rotation period, resulting in this phenomenon.

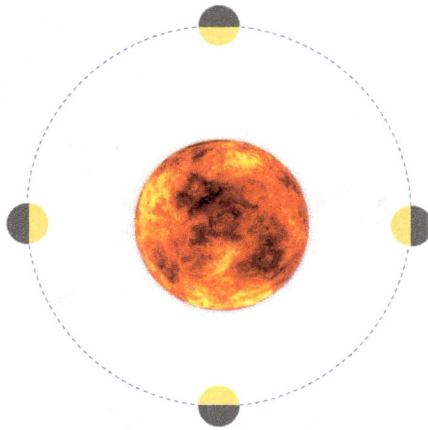

Fig. 5.11 A tidally locked planet will always show the same face to the parent star, resulting in one side being always day and another being always night. Credit: image by the author.

In the case of the TRAPPIST-1 system, due to the proximity at which the planets orbit the star, gravitational interactions cause the planets to only show the same side to the star. This means there is a side of the planet where it is always day, and another side where it is always night. This means that one side is always too hot and the other too cold to potentially sustain life. Interestingly, some scientists believe that the *terminator* of a planet, the dividing region between day and night (not the robot that tries to kill Sarah Connor and his son John), could be a good place to harbor life. Others think that as long as the planet has an atmosphere capable of redistributing heat around the planet, water could exist in a liquid form on either side and potentially sustain life.

The existence, or lack thereof, of an atmosphere on a planet plays an important role in the possibility of life on that planet. In the solar system, the conditions on Mars were extremely different in the past compared to what they are today. The Spirit and Opportunity rovers sent to Mars by NASA in 2004 and Curiosity and Perseverance sent in 2020, have found many features on the surface of the planet that indicate water was flowing on the surface of the red planet at some stage. Rivers, lakes, and oceans are believed to have been part of the Martian landscape.

Fig. 5.12 According to NASA's scientists, a very fast, deep river is the cause of these bands of rocks. Captured by NASA's Perseverance Mars rover between February 28 and March 9, 2023. Credit: NASA/JPL-Caltech/ASU/MSSS.

So, what happened? Why did Mars lose all of its water? The most likely explanation is because of its current thin atmosphere. NASA's Mars Atmosphere and Volatile Evolution (MAVEN), which was launched in November 2013 and arrived at Mars in September 2014, has found evidence that Mars' atmosphere has been escaping to space due to the effects of the Solar wind. MAVEN data indicate that gas is stripped away at a rate of 100 grams every second. This might not sound that much, but adding this up over billions of years results in an atmosphere that is so thin and cold today—leading to low atmospheric pressure— that prevents water in liquid form on its surface. In addition, Mars' mass is a lot lower than Earth's. We talked about the escape velocity in Chapter 2, and how the mass of the planet determines this. For a planet with low mass, it is harder to retain its atmosphere as all the gases will eventually escape to space.

As we can see, the thickness of the atmosphere is a very important feature for a planet to be able to support liquid water on its surface. This is why so many people, including me, got sad when astronomers reported in June 2023[13] that the existence of a thick carbon dioxide atmosphere for TRAPPIST-1 c was very unlikely. Using the JWST, and more specifically, its Mid-Infrared Instrument (MIRI), researchers determined that the dayside temperature of this planet is around 380 Kelvin (106.85 degrees Celsius), which does not favor the existence of a thick atmosphere on the planet. Astronomers collected lightcurves of the host star TRAPPIST-1 while the planet TRAPPIST-1 c transited in front of it on four different occasions. These lightcurves allowed the researchers to determine the brightness temperature of the planet. A similar result had been shared only three months before, in March 2023, for TRAPPIST-1 b.[14] TRAP-PIST-1 b is the closest planet to the ultracool dwarf star of the system. Also, in that occasion, by analyzing the star lightcurves collected during the transits of the planet around its host star, researchers were able to indicate that no detectable atmosphere of carbon dioxide was detected either. With day temperatures of around 534 Kelvin (260.85 degrees Celsius), astronomers have concluded TRAPPIST-1 b to have a very thin or maybe no atmosphere at all.

TRAPPIST-1 planets are all Earth-size planets, but there is a different type of planet that could be rocky but bigger than Earth, way bigger: The Super-Earths.

Super-Earths and Mini-Neptunes

Super-Earths are weird. We don't have those in the solar system either (or maybe the solar system is weird). These are planets between two and two and a half the size of Earth and between one to ten times Earth's mass. Let's mention one more time that our current techniques are biased in finding such large planets. Therefore, nearly 30% of all the planets astronomers have found belong to this category. As it is usual in astronomy (sometimes I think astronomers do it on purpose), the Super-Earth name is misleading as it does not necessarily mean that planets within this category are similar to our home planet or some sort of super-hero planet. Super-Earths can indeed be rocky, but they can also be made of gas, a combination of gas and rock, or they could even be water worlds like in the 1995 movie Waterworld with Kevin Costner (put this book down for now and go and watch it, it is pretty cool).

When a Super-Earth is found to be composed mostly of gas, astronomers call those *mini-Neptunes* or *sub-Neptunes*, as they are smaller than the planet Neptune in the solar system but have similar compositions.

The first Super-Earth discovered, GJ 876 d, is a planet with more than 7.5 times the mass of Earth. Its host star, GJ 876, is located at around 15 light-years from Earth. The planet, detected using the radial velocity method and reported in 2005,[15] completes a full revolution in less than two days. Like GJ 876 d, most Super-Earths astronomers have found planets that orbit very close to their host star.

Let's discuss a very interesting Super-Earth: Kepler-452 b.[16] This transiting planet exhibits the record for the largest semi-major axis— average distance to its host star— measured so far for a Super-Earth. Located at 1.046 AU of distance from the parent star, Kepler-452 b has a period of 384 days, which is a little bit longer than our terrestrial year.

Fig. 5.13 An artistic impression of Kepler-452 b, a likely rocky planet categorized as a super-earth, that orbits a Sun-like star located at 1,846 light-years from Earth. Credit: NASA.

Kepler-452 b, with 1.6 times the size of Earth but 3.3 Earth's mass, is considered by many as the most analogous planet to Earth. Apart from its orbiting period being very similar to Earth's orbital period, Kepler-452 b orbits a G class star, which is the same category the Sun belongs to. Like Earth, the planet is also located inside the star's habitable zone. Even more, scientists have a confidence of around 60% that this is a rocky planet. Given that Kepler-452 is a G star, the planet, which is at a similar distance from its star to the distance we are from the Sun, receives a similar amount of light and temperature. Even more, scientists believe the planet is about 6 billion years old, which means that, if there is life there, compared to life here on Earth, it would have had an extra 1.5 billion years to evolve. Definitely, food for thought.

We have discussed how astronomers classify exoplanets by its size and mass. Having a notion of how these characteristics compare to the planets in the solar system gives them a unique perspective when trying to understand how their respective planetary systems are configured.

Classification of Planetary Systems

The endeavor of classifying individual planets based on their physical characteristic serves one of the main goals that astronomers have: the classification of entire planetary systems. Achieving such a classification might help astronomers identify different formation mechanisms and provide insights into the dynamical history of planetary systems.

Many ideas have been proposed, but one of the most accepted and agreed ways of classification is based on the mass of the planets and the arrangement of those planets in their planetary systems. To this aim, four classes of planetary architectures have been suggested.[17,18]

Class I – Similar

In this type of planetary systems, all the planets have masses that are approximately similar to each other. These systems have been assigned the nickname peas systems as they resemble peas in a pod. Examples of such planetary systems include TRAPPIST-1 and its seven rocky planets and the six-planet system TOI-178 located at 204 light-years from Earth, with planets ranging from super-Earths to mini-Neptunes. These systems are the most common architecture. Observations indicate that out of all the planetary systems, the occurrence of Similar systems is approximately 58%.

Class II – Mixed

These are planetary systems where there does not exist a discernible regular pattern in the mass distribution of the planets regarding their distances to the parent star. The planets in these systems exhibit different ranges of masses, sizes, and distances between them. Examples of such systems include the GJ 876 system, located at 15 light-years from Earth. GJ 876 is an M-type dwarf star hosting four planets with masses between

8 to 888 Earth's masses. Another example of such systems is the one hosted by the star Kepler-89, located at 1,548 light-years from Earth. This system is comprised of four planets with masses between 10 to 100 times the mass of Earth. Only 5% of planetary systems have been observed to have this type of architecture.

Class III – Anti-Ordered

In these planetary systems, the mass of the planets decreases according to the distance from the star. Planets that are closer to the parent star display larger masses when compared to the ones farther away. These systems have only been reported in simulations as no observed planetary system has been found to follow this architecture. However, simulations indicate that up to 8% of all the planetary systems in the universe should exhibit this arrangement. The lack of evidence for such systems might indicate challenges in our current observing and detection capabilities.

Class IV – Ordered

This class is the opposite of the Anti-Ordered category. In this planetary system architecture, the mass of the planets increases according to their distance from the star. Closer planets have smaller masses, while farther away planets exhibit larger masses. This increase in mass could be regularly displayed in a consistent direction without any exceptions. For instance, in the planetary system TOI-561, located at a distance of 280 light-years from Earth, each of the four planets is consistently more massive than the one before it with respect to the distance to the star.

On the other hand, the increase in mass might follow a general trend, where an overall increase is observed, but some exceptions are present. Does that sound familiar? Yes, the solar system exhibits an ordered planetary architecture, with a general trend of planets increasing their masses with distance from the Sun, though some exceptions are present; for instance, Neptune, which is farther away from the Sun than Uranus, is actually less massive. Approximately 37% of observed planetary systems exhibit this arrangement.

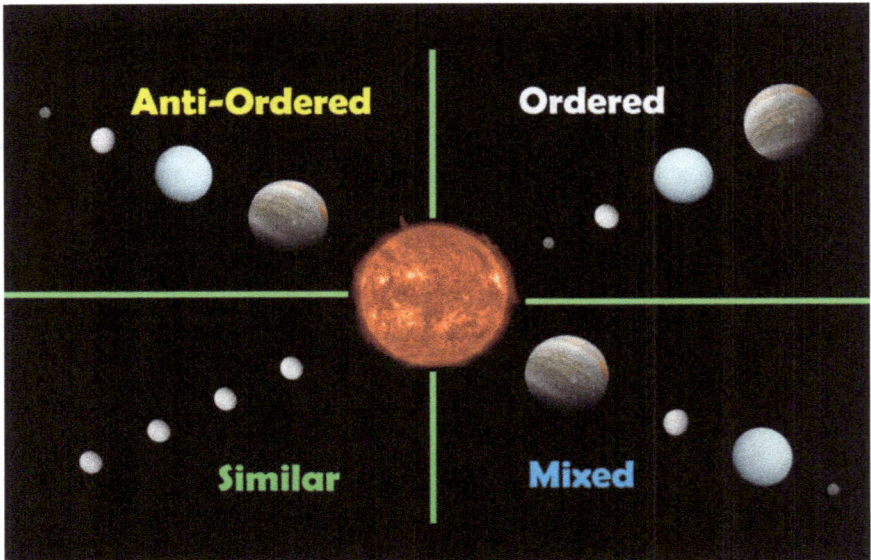

Fig. 5.14 Classes of planetary systems architectures. The arrangement of planets might help astronomers to understand the different mechanisms of planet formation. Credit: @AstroPhil2000.

We have explored how size and mass are very useful ways of classifying individual planets. The reader is invited to try the "Exodashboard" website.[19] This is a website I have put together using Phyton and Stream-lit.[20] Exodashboard imports the latest database from NASA's exoplanet archive, allowing the user to extract, export, and visualize interesting information about the exoplanets detected to date and their different classifications.

Besides mass and size, other useful types of classification involve the relation to how close planets are from their host stars, or what type of object they orbit, or even if they orbit anything at all! Let's explore other categories astronomers use.

Other Types of Classifications

Hot Jupiters

We have already discussed the existence of large planets that orbit so close to their stars that they complete a full orbit in a matter of days. It is worth mentioning one more time that current detection methods are heavily biased towards finding these big close-in planets as they affect the behavior of their host stars in a larger matter. In the case of Hot Jupiters, these are gas giant planets that orbit their parent star in less than 10 days. This means that these planets experience extreme irradiation and exhibit temperatures in the order of 1,500 Kelvin (1,226 degrees Celsius). Within the Hot Jupiters category are two subcategories: Very-Hot Jupiters (VHJ), which are planets that have periods between 2 to 3 days and orbit their stars as close as 0.1 AU. These planets have temperatures that usually are between 1,500 and 2,000 Kelvin (1,226 and 1,726 degrees Celsius). The second subcategory is the Ultra-Hot Jupiters (UHJ), which exhibit temperatures from 2,000 Kelvin to over 4,000 Kelvin (1,726 to 3726 degrees Celsius), usually at distances between 0.01 to 0.05 AU from their host stars. Ultra-Hot Jupiter planet's orbital periods vary between a few hours to less than a couple of days. The atmospheres of these planets resemble stellar atmospheres as the majority of their molecular constituents disassociate due to the extremely high temperatures to which they are exposed.

The fastest known UHJ, is the planet WASP-19 b which was discovered in 2009,[21] and orbits its parent star in only 18.9 hours. This means that what we consider a year on Earth lasts less than one of our days. The WASP-19 system is located at around 873 light-years from Earth, and the planet was discovered using the transit method.

Fig. 5.15 An artistic impression of WASP-19 b. The planet's atmosphere could be evaporating due to its close proximity to the parent star. Credit: image generated by OpenAI's DALL·E.

It is worth adding a small aside here. When astronomers talk about the temperature of a planet, they are referring to the *equilibrium temperature*. The equilibrium temperature is the temperature of a planet under equilibrium conditions, which means that a planet's total energy is constant. In other words, for practical purposes, it is assumed that all the energy received from their parent star is re-emitted to space, causing the planet's temperature to remain constant. This equilibrium temperature assumes that the planet's temperature is uniform over all of its surface. Earth's equilibrium temperature is estimated to be around -19 degrees Celsius (255 Kelvin). However, if the greenhouse effect is considered, a more pleasant 15 degrees Celsius (288 Kelvin) average temperature is estimated.

Of course, this is just an approximation and not the real deal. When I travel from Brisbane to Los Angeles in mid-December, I experience totally opposite seasons. When boarding the plane in Australia, Brisbane is experiencing the peak of summer, while Los Angeles is in the middle of winter. Therefore, my wardrobe choices need to change while onboard the plane; I have 12 hours to decide what to wear, so, no rush.

The equilibrium temperature of a planet is determined based on the average distance from the planet to its parent star— its semimajor axis, the size and temperature of the parent star, and the *albedo* of the planet, which describes how much light is reflected back to space by the planet's surface. However, the equilibrium temperature is not the same as a planet's surface temperature, which in the end, is the one that determines the habitability of the planet. Earth's equilibrium temperature of -19 degrees Celsius is obviously below the freezing point of water. Fortunately for life on Earth, this is not the planet's temperature at its surface. To calculate the temperature at the surface of a planet, it is necessary to consider models that take into account the atmosphere of the planet. On Earth, the atmosphere is responsible for a large part of the *greenhouse effect*, increasing the global temperature and allowing water to exist in its liquid form on the surface. Astronomers, therefore, are desperate to get their hands on data that can help them characterize the atmosphere of exoplanets as this would allow them to estimate the surface temperature of those planets. Thankfully, this data is what the JWST has started to provide.

Although the equilibrium temperature is not a perfect metric for defining a planet's habitability, it remains an important piece of the puzzle. For planets situated extremely close to their stars and with either very thin or non-existent atmospheres, the equilibrium temperature can be used as a reliable indicator of the planet's actual temperature.

Hot Jupiters are extremely close to their host star. But how close can a planet get to their host star before being torn apart? There is a limit known as the *Roche Limit*, named after Edouard Roche (1820-1883). The Roche Limit, determined by the density of the bodies involved, is the minimum distance a planet can get close to its star without being destroyed by its gravitational tidal forces. Beyond such a distance, the differential gravitational force (stronger on the side facing the star and weaker on the far side) causes the planet's shape to become increasingly elongated, causing oscillations all across the planet, with the end result of the planet coming apart.

The Roche Limit is not only relevant to planets around stars but also to moons around planets. One of the most popular explanations of the origin of Saturn rings indicates that these rings are the result of a moon or moons getting too close to the planet. The total mass of Saturn's rings is comparable to that of one of the Saturnian mid-sized satellites, like Mimas. Other explanations for the origin of Saturn's rings include left-overs of Saturn's planet formation, pieces of a comet that felt adventurous and got too close to the gas giant, or just pieces of multiple small moons that collided with one another due to the gravitational effects of the larger ones and the planet itself.[22] A strong hypothesis is presented on a recent research[23]. The authors of this paper report that two icy moons could have collided and shattered a few hundred million years ago. The impact would have produced a wide distribution of debris, including large chunks of pure-ice material that entered Saturn's Roche limit. This event could have led to either the formation or the rejuvenation of Saturn's rings system. In addition, the impact could have scattered debris throughout the system, causing disturbances and collisions with other bodies, contributing to the formation or maintenance of the rings.

Fig. 5.16 A view of Saturn's rings system as imaged by the Hubble Space Telescope. The Roche Limit could have played an important role in the formation of the rings. Credit: NASA.

On a similar note, astronomers have estimated that the Mars' moon Phobos may potentially be destroyed in a not-too-distant (in universe scales) future as it gets too close to the planet. Estimations indicate that Phobos will reach its Roche Limit within 20 to 40 million years,[24] at which point it will disintegrate, leaving behind a nice small ring system around Mars.[25]

Fig. 5.17 An astonishing image of Phobos, by ESA's Mars Express. The moon is orbiting at only 6,000 km over Mars' surface and dangerously approaching to the planet's Roche Limit. Credit: ESA/DLR/FUBerlin/AndreaLuck CC BY and colorized by Andrew Luck.

Planets that orbit two or more stars

We briefly discussed circumbinary planets when we introduced the exoplanet detection method Eclipse Timing Variations (ETV) in Chapter 4. Essentially, a circumbinary planet is a planet that orbits a binary star system. As weird as this sounds, binary star systems seem to be quite common in the universe. Astronomers estimate that about 85% of all the stars have a companion or multiple companions.

Beyond two stars, systems with three or more stars orbiting around each other are called *multiple-star systems*; planets that orbit a three-star system are known as *circumtrinary* planets. More generically speaking, I was surprised that I could not find a general definition for a planet that orbits a multi-stellar system, so I decided on the *circummultinary planet* to describe those. Although being catchy, I must admit that perhaps *multi-stellar planet* is a more appropriate and easier-to-pronounce term.

A well-known example of a multiple-stellar system is the Alpha Centauri planetary system, the closest stellar system to the Sun. The Alpha Centauri system is located at only 4.37 light-years away and is comprised of three stars, Alpha Centauri A, also known as Rigil Kentaurus, Alpha Centauri B or Toliman, and Proxima Centauri, the closest star to the Sun and the star in this system that appears to have a planet orbiting around it.

Proxima Centauri is a red dwarf star that has a luminosity of only 0.15 percent of the Sun, along with a radius that is merely 14 percent of the Sun's radius and a mass of about 12 percent of the mass of the Sun. The planet, Proxima Centauri b, or just Proxima b, discovered in 2016 using the radial velocity method,[26] completes a full orbit around the red dwarf star in only 11.2 days, and it is located at just 0.05 astronomical units. This planet is not only interesting for being the closest exoplanet to Earth, but also because, according to recent discoveries, it seems to be in the star's habitable zone.[27] Technically speaking, Proxima b orbits only a single star. However, since Proxima Centauri orbits the other two stars in the system, Proxima b can safely be categorized as a multi-stellar planet.

Nature abounds in contrasts; while planets like Proxima b exist within a multi-stellar system, there are also some planets that do not belong to any planetary system. Let's talk about these fascinating outliers.

Rogue Planets

Planetary systems that are still forming are chaotic places. Planetesimals, moons, and recently formed planets resemble a cosmic billiard table where they collide against each other. These collisions can cause certain celestial objects to be set free from the gravitational influence of the central star. Every now and then, planets get hit by other planets or objects so hard as to cause them to escape the gravitational influence of their parent star. These planets are now free to roam the universe without being gravitationally bound to a host star. Commonly known as *Free-floating* planets, they are also dramatically referred to as *Rogue planets*.

In the 'Star Trek: Enterprise' episode, "Rogue Planet" (one of my favorite episodes in this franchise), Lieutenant Reed encounters one of these celestial bodies. Apparently, the planet had broken out of orbit from its host star and just roams the universe freely without any strings attached. As the planet does not have a parent star and, therefore, does not receive any form of light, the crew of the Enterprise believe that the planet must be devoid of any life. This results not to be true, but I won't spoil it for you in case you want to check it out.

The episode correctly states that light does not reach its surface because the planet does not orbit any star. This situation raises a few interesting questions. First, are such objects even considered actual planets? Let's remember that according to the International Astronomical Union (IAU), the first requirement for a celestial object to be categorized as a planet, is that the body should orbit a star. Therefore, rogue planets should not be considered planets, right? The second question is, how do you even detect such an object? The most common techniques, radial velocity and transit, indirectly detect a planet by measuring their influence on their parent star; no parent star, no influence to measure. However, rogue planets still have mass, and large ones, even as large as five times the mass of Jupiter. This means that these objects' masses will bend light and, therefore, can be detected using the *Microlensing* technique discussed in Chapter 4.

As a matter of fact, for a long time, Gravitational Microlensing was the only way to detect these unbound planets. Let's remember that in a microlensing event, a foreground object is detected as it magnifies the light coming from a background source. Taking advantage of this, the groups Microlensing Observations in Astrophysics (MOA) and Optical Gravitational Lensing Experiment (OGLE), conducted microlensing surveys toward the central region of our galaxy, also known as the *bulge*. The collected data included the light curves of 50 million stars in the bulge of the galaxy at frequent intervals of time, ranging between 10 to 15 minutes. The analysis revealed brief microlensing events lasting less than two days. These observations led to the discovery of ten events caused by planetary-mass lenses.[28] While this is certainly an impressive achievement, gravitational microlensing events are rare, and the probability of everything working out in a way that astronomers can detect such events is relatively low. Therefore, it is better to have a backup plan.

A second way to find rogue planets is to take advantage of their characteristics when they are young. Like their still-gravitationally-bound-to-a-star counterparts, young rogue planets are also very visible in infrared. Therefore, a group of researchers took advantage of this fact and examined 80,818 individual wide-field images collected using 18 different cameras over the last 20 years. [29] The resulting catalog, named Dynamical Analysis of Nearby Clusters, or DANCe, was used to find new 70 free-floating planets.

However, it appears that some free-floating planets might not be destined to spend their lives in total isolation. Surprisingly, a type of free-floating planet is lucky enough to have been ejected in pairs. Scientists from the European Space Research and Technology Centre (ESTEC), announced in September 2023 the discovery of what they call Jupiter Mass Binary Objects (JuMBOs).[30] Using a new JWST infrared survey of the inner Orion Nebula and Trapezium Cluster, both located roughly 1,344 light-years from Earth, researchers have discovered and characterized 540 planetary-mass rogue candidates. The Orion nebula is one of the closest regions of massive star formation, while the Trapezium Cluster is composed of several hot, young stars, making them good targets for finding young planetary systems. Out of all the identified planetary-mass rogue candidates, 49 of those objects are in a binary configuration. This means these double-planets were ejected in pairs and somehow stuck together. Such a finding is completely unexpected, and there are no current models of planetary system formation that predicted the ejection of planets in a binary configuration. These binaries are young planets with surface temperatures of around 1,273 Kelvin (1,000 degrees Celsius) but will quickly cool down given that they don't orbit any star.

All the planets detected thus far are within our galactic neighborhood. Due to the current limitations of our detection techniques, planets within our galaxy are easier to find. However, there is nothing in the current models of planet formation to suggest that stars in other galaxies cannot host planets. Such planets are referred to as *extragalactic planets*.

Extragalactic planets

Astronomers estimate at least 100 billion galaxies in the observable universe. Astronomers also estimate that, on average, each galaxy contains 100 billion stars. So, if we want a rough estimate of the number of stars in the universe, we need to multiply the number of galaxies in the universe (100 billion) by the number of stars per galaxy (100 billion). The result is a number with a one followed by twenty-two zeros, which expressed in scientific notation is 1×10^{22}, or ten sextillion stars. Each star can potentially host multiple planets, like the Sun, with eight planets. Others may have no planets, or some may have more or fewer than eight. To simplify calculations, astronomers assume that, on average, every star has at least one planet. Therefore, if we assume that every star has, on average, one planet, we can estimate the number of planets in the universe to be at least equal to the estimated number of stars. That would be ten sextillion planets! If we further assume an average of two planets per star, the number of planets in the universe would double, and so on.

With these numbers, it is only expected to find at some stage, planets that are not within our own galaxy. A group of researchers reported one of these possible planets in 2021.[31] The candidate planet is believed to be orbiting the X-ray binary M51-ULS-1 in the galaxy M51. An X-ray binary is a system where a stellar remnant (known as the accretor), which could be a neutron star, a white dwarf, or a black hole, accretes matter from a companion star. As the stolen material from the companion star falls onto the stellar remnant, it becomes a strong source of X-rays. Similar to the transit method, where a dip in the brightness of a host star is observed when a planet passes in front of it, a planet orbiting the accretor in an X-ray binary system could cause detectable X-ray eclipses. These X-ray eclipses are what the researchers of the 2021 paper reported.

An advantage of X-ray transits over optical light transits is that X-rays are concentrated in a small area. Consequently, a transiting planet can block a significant portion, or even all, of the emitted X-ray light. This makes X-ray transits more detectable at longer distances compared to visible light transits. However, if a planet is indeed causing the detected X-ray transits, the chances of finding life there are low. The presence of the stellar remnant indicates that a star died in the past, releasing enormous amounts of radiation that could have affected any planets in the system, and more than likely, eliminating any possibility of life.

This is life as we know it. However, how can we discover forms of life that are unknown to us? After all, we can hypothesize about different chemistries, but sticking to what we know is the easiest. This is why researchers' search efforts are primarily directed toward finding life like that which inhabits the Earth.

In the next chapter, we will examine the efforts of scientists to find such a life and discuss potential techniques that may be used in the future. Don't go away. I promise, things are getting even more interesting.

Chapter 6
Looking for the Signatures of Life

> "Those are primes! 2,3,5,7, those are all prime numbers and there's no way that's a natural phenomenon!"
>
> — Contact, Eleanor Arroway (1997)

TL;DR

With potentially ten sextillion planets in the universe, the possibility of finding life beyond Earth seems very promising. Celestial beings have been part of our culture since the dawn of time. Millions of people believe that we have, and we continue to be, visited by extraterrestrial beings. The formerly Unidentified Flying Object (UFO) phenomenon, now rebranded as Unidentified Aerial Phenomenon (UAP), is as relevant as it has been in the last 60 years. Despite all the claims and witnesses, no concrete evidence or proof has been made available.

Astronomers and researchers are not holding their breath while waiting for such a proof, and they continue to study and search the universe for a signal that can be used as evidence that we are not alone.

In the search of extraterrestrial life, researchers look for a signal that is detectable and measurable and that could indicate the current or past presence of life. A signal like that is known as a *biosignature*, and can be created by non-intelligent and intelligent life alike. However, intelligent life could have developed tools and eventually technology, like humans on Earth. Therefore, a signal that is a manifestation of their technological capability could, in theory, be detected. These signals are known as *technosignatures* and are defined by Dr. Jill Tarter as "evidence of some technology that modifies its environment in detectable ways".

Scientists search for biosignatures either *in-situ* or *remotely*. An In-situ search requires a probe to be deployed into the celestial object of interest and the collection of samples from the environment (another planet, a moon, or an asteroid) to be analyzed, either by the instruments on board or in specialized labs which requires the transportation of the samples back to Earth; at least for now.

For instance, the Perseverance rover, or Percy, is the latest mission on Mars. Percy is currently collecting samples and analyzing them with its onboard instruments. However, NASA and ESA are planning a mission that will bring those samples back to Earth by the early to Mid-2030s.

Due to the limitations in our current propulsion technologies, in-situ search is only possible for certain objects in the solar system. However, for more distant bodies in the solar system and exoplanets in general, scientists are only able to search for life remotely. Techniques such as spectroscopy and chromatography help scientists detect traces of chemical elements such as methane and carbon dioxide, which are commonly associated with metabolic processes caused by life here on Earth.

Similar to biosignatures, technosignatures can also be detected in-situ and remotely. Technological artifacts could have been left behind by an extraterrestrial civilization during one of their exploration missions to the solar system, like humans have on Mars and the Moon. As a matter of fact, a very small number of astronomers (this number continues to reduce as time goes by) believe this was the case of Oumuamua, the first interstellar object that astronomers have a record of.

Remote detection of technosignatures usually involves the detection of radio or optical signals from outer space. Coming from our own experience and technology, such signals are probably narrowband signals, given that these are easier to transmit and tune by a receiver.

Technosignatures could be intentional or the by-product of an alien civilization's existence. It is thought that a technologically advanced alien civilization will produce large amounts of energy which astronomers could detect. This energy consumption concept is explored in the Kardashev scale, proposed by the Soviet astronomer Nikolai Kardashev in 1964. The Kardashev scale describes three types of civilizations: *Type-I*, which is a civilization that has learned to efficiently utilize all the resources of their home planet. *Type-II* is a civilization that can extract all the energy produced by its parent star, and *Type-III* civilization is a civilization that can harness the energy produced by its home galaxy. With this in mind, a type-II civilization, for instance, could have developed a Dyson sphere, a megastructure around a star, which could potentially be detected due to the excess of heat (infrared energy) radiated.

With the aim of finding non-human transmitted signals, the Search for Extraterrestrial Intelligence (SETI) institute was established in 1984. However, the discussion had started two decades earlier, in 1961, on a seminar that took place at the Green Bank observatory in West Virginia. The seminar was planned by the astronomer Frank Drake, and several distinguished scientists attended. The seminar was part of project Ozma, which is considered the first modern attempt to detect interstellar transmissions.

SETI professionals and amateurs alike have been looking for extraterrestrial signals for over 60 years. The search has mostly focused on radio waves, given that these are ideal for long-range communications. Radio waves are less susceptible to be scattered or absorbed by the dust that plagues interstellar space. The favorite frequency astronomers tune their receivers to is the 1,420 MHz frequency, as this is the frequency of the

21-centimeter emission line of interstellar hydrogen. It is believed that the knowledge of such a frequency is shared by any technologically mature society out there and considered very likely that they decide to transmit in this frequency.

As resources are very limited, researchers must come up with new ideas to maximize the search. Dr. Sheikh, for instance, proposed the concept of *axes of merit*. Dr. Sheikh's work establishes a framework in which searches for technosignature are objectively and quantitatively evaluated to identify which ones astronomers' efforts should be focused on.

The search has been going on for a while, with no results of an alien civilization to show so far. However, SETI professionals have only searched a fraction of the universe, roughly equivalent to the ratio of a large hot tub to that of Earth's Oceans.

In all these years of searching, the most promising candidate signal is known as the Wow! signal, detected in August 1977. This signal has all the right characteristics of a technosignature and has not been totally discarded.

SETI initiatives are reactive, which means that astronomers limit themselves to just "listen" patiently. On the contrary, initiatives such as Messaging Extraterrestrial Intelligence (METI) propose sending an intentional message to the stars to let the rest of the universe know about our presence. METI has been done by including records and plaques on several spacecraft that have left the solar system and by virtue of sending targeted messages to the globular star cluster M13 located at 25,000 light-years from Earth. Detractors of such a strategy are afraid of humanity running into a similar fate to the ones faced by less technologically advanced civilizations meeting a more advanced counterpart here on Earth. History is full of such examples.

In the meantime, recent advances in Artificial Intelligence (AI) and Machine Learning (ML) techniques are assisting researchers in searching for signals that could indicate that we are not alone.

Fig. 6.1 The Allen Telescope Array in California is one of the facilities that is actively used for SETI searches. Credit: Astronomy Staff.

Angels among us

Entities from the sky have captivated the imagination of humanity since the dawn of time. Almost every culture on Earth has stories about celestial beings coming from the heavens. These celestial beings are usually portrayed with human-like features but are distinguished by wings to signify that they come from above. We have called these beings angels in the past, but nowadays we call them aliens. I am not the first person who has noticed this. The Canadian poet Margaret Atwood, author of "The Handmaid's tale", argues that: "Extraterrestrials have taken the place of angels, demons, fairies, and saints...".[1]

In July 2023, the US congress hosted a public hearing about Unidentified Flying Objects (UFOs) or, –how it is referred to nowadays– Unidentified Aerial Phenomenon (UAPs). The hearings aimed to explore what the US government knows about UAPs and the implications of those objects as a matter of national security. Potential key witnesses, or what the media calls them, "whistleblowers", claim that UAPs or UFOs, whatever you want to call them, are indeed extraterrestrial in nature –notice that the acronyms don't imply that this should be the case– and that the Pentagon is in possession of several spaceships that apparently have crashed on our planet. They also claim that, from those crash scenes, the US government has been able to recover bodies or even survivors, alien survivors.

I am not going to discuss the validity of these assertions as my knowledge of the internal politics of the USA is almost non-existent, and the focus of this book is on available scientific evidence. Given that science findings are thoroughly scrutinized and openly shared, I can confidently say there is no concealment of information regarding beings visiting our world.

Finally, it seems quite odd to me that an extraterrestrial civilization that has been able to master the intricacies and physics of interstellar travel manages to crash their spaceships when reaching Earth, and not just once but many times in a relatively short period of time. Perhaps all these crashes are the consequence of inexperienced alien teenage pilots still on their learning driver's license who stole their parents' ships and decided to visit Earth?

Fig. 6.2 Are alien teenagers stealing their parents' spaceships and crashing on Earth? Credit: image generated by OpenAI's DALL·E.

In the meantime, astronomers continue searching for signs of extraterrestrial life, specifically for signals or signatures that astronomers can detect and analyze. The branch of astronomy that deals with extraterrestrial life is called *Astrobiology*. In essence, life generates by-products as a consequence of their existence. At home, for instance, we might recognize that we are not alone if we hear someone or something breathing in another room. The person or animal with us at home doesn't intend to communicate their presence to us, but they inadvertently let their presence be known just by being alive; this is an example of *biosignature*. More formally, a biosignature is a signal that is detectable and measurable, and that could indicate the current or past presence of life. On the other hand, we might infer that we are not alone at home because we suddenly start hearing noises resembling a narrator relating to a football game. This might indicate that someone has turned the TV on. Like in the previous example, the person in the next room was not trying to let us know about their presence, but by us hearing the sounds coming from the TV and leaving any paranormal activity aside or a Google assistant pre-programmed routine, we can be sure that someone is indeed in that room. In this case, we detect the presence of that someone because they are using technology, and we can detect the physical manifestation of it. This type of signal is called *technosignature* as it is a signal derived from a technological agent's existence. Notice that technosignatures are a subset of biosignatures. As such, potential signals indicate the existence of past or present life. Let's talk about biosignatures and technosignatures in more detail.

Biosignatures

Before discussing how scientists can potentially detect life, we need to start by defining life itself. This task is not as straightforward as it might seem. Defining life carries profound implications from both scientific and philosophical perspectives. Full disclaimer: scientists have not yet reached a consensus on a definition for life.

Unfortunately, scientists currently have access to only one sample of life in the entire universe: life on Earth. This, which is referred to as the *biogeocentric* paradigm or vision,[2] introduces a bias in the search for life in the universe. To avoid such a bias, a general definition of life is required. For instance, NASA's definition of life reads: "Life is a self-sustaining chemical system capable of Darwinian evolution".[3, 4] The beauty of this definition is that it is not only intended to be applied to life as we know it, or *terran* life, but also to cover the definition of any potential life elsewhere. However, Earth remains our only reference point for inhabited planets.

From this definition, we can see that life is considered to be chemical. Living organisms perform chemical transformations to take advantage of the material surrounding them to support and prolong their existence.

Fortunately, as life transforms, or more technically, metabolizes, surrounding material to convert nutrients into energy through chemical reactions, it produces signatures that can be measured and quantified.

Through biological processes, living organisms produce signatures that can be very simple but also extremely complex. However, issues arise when non-biological processes can also produce biosignature-type signals, creating so-called false positives. In essence, scientists may think that they have found proof of life, but in reality, they have not. A typical example is the oxygen present in the Earth's atmosphere. On Earth, most oxygen is produced by photosynthesis, mostly by oceanic plankton and terrestrial plants. Plants convert sunlight, water, and carbon dioxide into oxygen. The geological activity would cause oxygen to disappear rapidly for an Earth-like planet. Therefore, on Earth, oxygen is a clear indication of the presence of life. In other words, oxygen is a strong biosignature. Plants produce oxygen, rapidly replenishing our atmosphere with this chemical element. However, oxygen can also be produced by the non-biological process known as *photo-disassociation* of water. Photo-disassociation is a process in which ultraviolet light can break down water molecules into their constituent elements, hydrogen and oxygen. Therefore, the sole presence of these chemical elements in a planet's atmosphere does not mean that life is responsible for it. Therefore, the detection of oxygen can be categorized in many cases as a false positive for life.

Another clear example of a potential false positive is phosphine. Phosphine on Earth can be produced by both non-related to life (abiotic) or related to life (biotic) processes. Certain bacteria can produce phosphine during their metabolic process in oxygen-deprived environments. However, volcanic activity can also produce phosphine gas when phosphorous-rich volcanic rocks come into contact with water. In 2021, a group of researchers published an article[5] in which they claimed to have detected phosphine in the clouds of Venus. In this paper, the authors discuss how the detected phosphine cannot be explained by any abiotic process on the planet. Therefore, the possibility of biological production remains a strong candidate for explaining the presence of phosphine. However, Carl Sagan once said that "extraordinary claims require extraordinary evidence". A second group of researchers published another article[6] in which the results of the first paper are questioned. By performing a re-analysis of the data, the authors of the second article concluded that the first group's methodology gave them a false positive. The controversy continues.

In this case, the detection of phosphine was done *remotely*, analyzing the spectrum of the light collected from the planet Venus.

The alternative to remotely detecting life is *in-situ* collection and analysis. For instance, sending a robot to Venus to take a sample from its clouds would allow scientists to perhaps set the controversy once and for all.

In-situ Analysis

The public is quite familiar with in-situ collection and analysis on Mars, where many rovers have traveled and collected samples. Five missions have been sent to Mars: Sojourner in 1996, Spirit and Opportunity in 2003, Curiosity in 2011, and the most recent, Perseverance in 2020. At the moment of this writing, Perseverance, nicknamed Percy, has been on Mars for 872 Sols. A Sol is the term used to represent a Martian day which is approximately 24 hours and 39 mins. According to the mission's NASA's website, Percy's mission is to "seek signs of ancient life and collect samples of rock and regolith (broken rock and soil) for a possible return to Earth". Percy has several instruments that allow it to perform in-situ analysis of the collected samples. However, NASA's and ESA's (European Space Agency) idea is to send another mission to collect those samples and bring them to Earth to perform more detailed analysis in advanced laboratories. The whole thing looks like it has been taken from a science fiction book. The plan is to send a vehicle to Mars in 2028. The vehicle, named Sample Retrieval Lander (SRL), will meet Perseverance at a predefined location. Perseverance will then load the collected samples into the SRL. Two small helicopters will provide additional capability to retrieve samples. Once all the samples have been loaded, they will be launched off from Mars on board the Mars Ascent Vehicle (MAV) rocket, which will meet the Earth Return Orbiter (ERO) space-craft in Mars' orbit. The ERO spacecraft will bring the samples to Earth safely in the early to mid-2030s. Once the samples are here on Earth, state-of-the-art labs will be used to analyze them.

Sample Retrieval Lander   Mars Ascent Vehicle   Earth Return Orbiter

Fig. 6.3 The different vehicles involved in NASA's and ESA's Mars samples retrieval mission. Credit: NASA.

Fig. 6.4 Multiple robots will team up to retrieve the samples collected by the Perseverance Rover to bring them safely and securely to Earth. Credit: NASA.

For obvious reasons, in the field of exoplanets, in-situ analysis is out of the question. The closest potential planet to Earth is four light-years away, and a probe with our current technology would take close to 18,000 years to travel that far. Recent technological proposals such as the

Starshot initiative might reduce that time considerably. Starshot aims to employ light-powered space travel to achieve speeds that are a significant fraction of the speed of light.

The idea is to have ground-based light beamer facilities that will push small space probes with lightsails attached to them. These very small space probes, or *nanocrafts*, could potentially reach the closest star planet to Earth in just 20 years—which is way better than 18,000 years and within the lifespan of a human being.

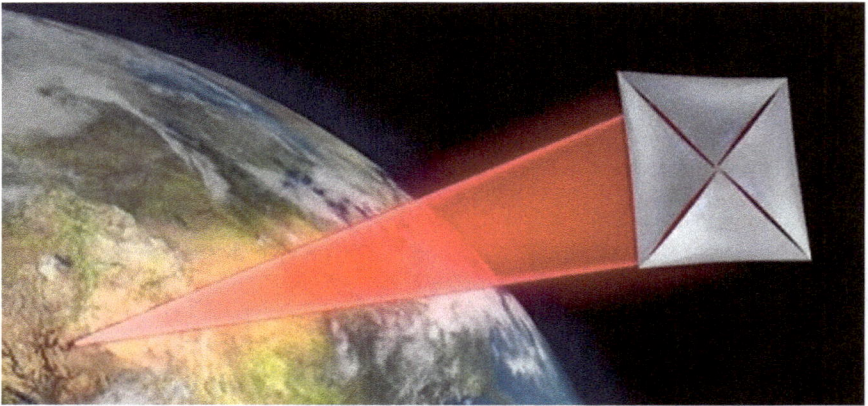

Fig. 6.5 A nanocraft is pushed by a light beam from a ground-based facility on Earth. Credit: Sky & Telescope.

Remote Analysis

While scientists and engineers figure out how to reach the closest star, astrobiologists still need to have a day (or night?) job and continue to remotely search for biosignatures. To this aim, they employ spectroscopy —the breaking up of light into its component colors using a prism— and photometry—the measurement of the brightness of a star in an image, two well-known astronomical techniques.

There are two methods astronomers use to "capture" and analyze a planet's atmosphere. In the first method, the reflected light coming from the planet itself is isolated from the light produced by the star. As observed in Figure 6.6, astronomers capture the combined spectrum from the star and the planet when the planet is in front of the star. Once the planet goes behind the star and, therefore, becomes invisible, astronomers are only able to obtain the light coming from the star. By suppressing the light collected for the star from the light collected for the star and the planet, astronomers isolate the light coming directly from the planet.

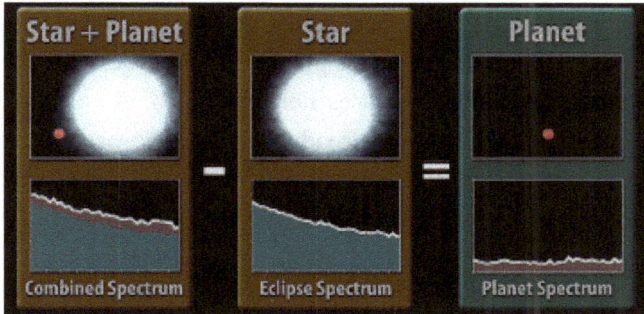

Fig. 6.6 The spectrum of a planet can be isolated once it vanishes from our point of view. Credit: NASA/JPL-Caltech/R. Hurt (SSC/Caltech).

The second method analyses the starlight when passing through an exoplanet's atmosphere. Gases in the planet's atmosphere will block certain parts of the spectrum in the starlight. This means that these gases have absorbed slices of the spectrum, indicating the presence of the gas in the exoplanet's atmosphere.

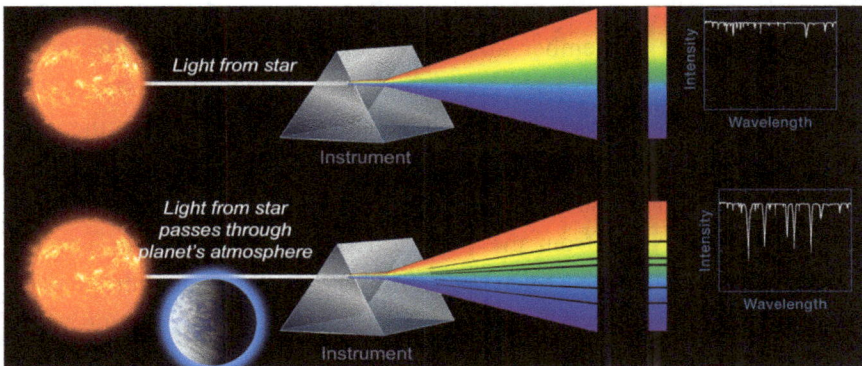

Fig. 6.7 Gases in the atmosphere of an exoplanet will cause black bands in the starlight spectrum as it traverses the planet's atmosphere. Credit: NASA/JPL-Caltech.

In the search for biosignatures, scientists deal with two types. First, there are the biosignatures we know of, and second, the ones we don't know of. The first ones are the signatures we are familiar with as they are the signatures produced by life on earth. We know of certain gases that are made up of chemical elements and molecules with well-known spectra. This is why instruments on board the JWST are extremely important in the search for life. They can detect, for instance, methane and carbon dioxide in the atmospheres of other planets. These chemicals are usually associated with metabolic processes here on Earth. As a matter of fact, scientists using the JWST have already detected the presence of carbon dioxide on a distant planet.[7] Astronomers made this amazing discovery by analyzing the spectra collected from WASP-39 b, a planet that orbits a main sequence star on a 4-day period. The planet itself is too close to the parent star, with an orbit of only 0.05 AU, and therefore, too hot to harbor any life. This means that the presence of carbon dioxide is not the result of life. One of the hypotheses for the presence of such a component is that, possibly, comets and asteroids bombarded WASP-39 b during its formation period, bringing carbon and oxygen to its surface; something that scientists believe happened during the Earth's early days. Despite this detection not indicating life per se in this case, it does help to demonstrate the capabilities of JWST in detecting biosignatures, which is extremely exciting and a big step in the right direction in our quest to find life elsewhere.

Researchers at the University of Cambridge announced a second important discovery in September 2023.[8] By analyzing spectra collected by JWST from the K2-18 b planet, which orbits the cool dwarf star K2-18, they discovered the presence of methane and carbon dioxide in the planet's atmosphere. The difference with WASP-39 b is that K2-18 b is in the habitable zone of its star, and recent studies have even suggested that could it be a *hycean* (from the words "hydrogen" and "ocean") world. These observations suggest that the world potentially has a hydrogen-rich atmosphere and a surface covered by a large ocean (again, go and watch that Kevin Costner movie).

K2-18 b, considered a Super-Earth and discovered in 2015 as part of the efforts of the extended Kepler mission K2, orbits its parent star on a 33-

day orbit, and its mass has been estimated to be 8.6 times as massive as Earth. The K2-18 system is located at approximately 120 light-years from Earth and the host star is colder and smaller than the Sun.

The team was able to characterize the atmosphere of the planet by analyzing light from the host star when passing through the exoplanet atmosphere. In addition to the carbon dioxide and methane found, their analysis also suggests the potential signs of the dimethyl sulfide (DMS) molecule. Such a discovery is a very interesting one, given that, in the oceans of Earth, even though DMS can be produced by abiotic processes like sunlight and chemical reactions, the production of this molecule is mostly attributed to biological transformations, particularly bacterial production and consumption.

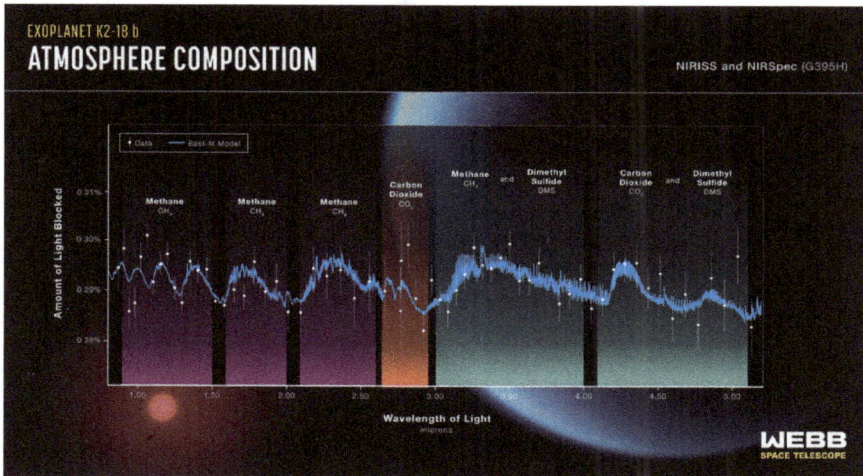

Fig. 6.8 Spectrum analysis of the atmosphere of K2-18 b. This is the first time that carbon dioxide, methane, and DMS have been detected in the atmosphere of a habitable zone planet. Credit: Illustration: NASA, CSA, ESA, R. Crawford (STScI), J. Olmsted (STScI), Science: N. Madhusudhan (Cambridge University).

Essential Conditions for Life

Apart from certain gases in the atmosphere that can be attributed to biotic processes, there are certain characteristics in a planet that can give scientists some hints about the existence of life. Here on Earth, one of these characteristics is pigmentation. Life can modify the landscape of a planet at a level that could be detected from outer space.[9] Vegetation on Earth, for instance, "paints" parts of the Earth's surface with a distinctive green color that could be identified from space. In addition, living organisms produce patterns, structures, and textures that can be recognized, and certain minerals that could not exist without the direct intervention of a living organism.

Unfortunately, efforts to find life elsewhere have yielded no results thus far. But the universe is huge, and we have only scratched the tip of the iceberg with the more than 5,000 exoplanets found so far. How do researchers know which part of the cosmos they need to focus their efforts on and what targets they need to point their instruments to? Researchers agree in four fundamental concepts[10] or requirements for life to occur. First, life needs a thermodynamic disequilibrium environment. This is a lot of fancy words. A thermodynamic disequilibrium environment is one where there is an imbalance of energy, matter, or chemical potential that allows processes to occur within the system. In the early Earth, for instance, it is hypothesized that simple molecules were produced in the atmosphere because of lighting or photochemistry in the crust by water-rock reactions or by effects of water on carbon meteorites that crashed on the planet. These simple molecules then reacted with other elements on Earth, such as water and minerals, to produce even more complex molecules. Therefore, an environment that allows for such a transfer and conversion of energy is critical to the potential emerging and success of life. A planet with an atmosphere that contains waste gases from living organisms with sufficient fluxes that alter the planet's reflectivity or albedo is a good candidate to explore. However, as with other possible biosignatures, the existence of thermodynamic disequilibrium does not necessarily relate to the presence of life; it may be crucial for it but not a consequence.

All planets by definition orbit a star, providing them with energy flux from sunlight. Even more, planets may have internal geological processes, such as volcanism and tectonic plates, that can provide extra energy. This is why researchers intend to compare Earth's disequilibrium, which we know is due to living organisms, with the disequilibrium observed of other planets in the solar system, which we know are due to abiotic processes.[11] In doing so, researchers aim to understand the difference so they can apply that filter to the atmospheres of exoplanets.

Second, life is complex, and needs complex molecules to be formed. Therefore, an environment that facilitates such complexity to arise is essential. Researchers talk about an environment capable of maintaining *covalent bonds*, particularly between atoms of carbon, hydrogen, and other atoms. A covalent bond is a chemical bond in which atoms share electrons, resulting in stable molecules. This type of bond is the most common form of chemical bond in living organisms. Covalent bonding allows molecules to share electrons with other molecules, facilitating the creation of long chains of compounds. This results in greater complexity in life. Complexity is important because you need a way to transmit and encode crucial and specialized information. Information is needed so a living organism can diversify, adapt, and interact with its surroundings, enabling it to thrive in a range of environments and perform a wide range of functions. Covalent bonds are essential to form carbon-based organic molecules, like our DNA and proteins, which, without them, we would not have life as we know it here on Earth.

Covalent bonds formation requires environments that facilitate interactions between atoms and where electrons can be shared easily. In addition, they need a suitable solvent like water, allowing reactant substances to be mixed. This is the third element for life to prosper: The existence of a liquid environment. On Earth, water provides a medium for reactions to occur. This is why researchers focus their efforts on finding planets where liquid water could exist on their surfaces. The term habitable zone or *Goldilocks zone,* refers to a zone where planets orbit at a distance from their parent star so that water can flow on the planet's surface. The temperatures experienced by a planet too close to its host star can be even greater than some stars' surface temperature, causing any potential liquid water to evaporate. On the other hand, if the planet is too far away from the parent star, water will be frozen.

In addition, for a planet that is too close to its host star, extreme pressures are also experienced, which is not good for covalent bonds to form, not only because any possible solvent may be evaporated but also because such temperatures and pressures would not allow electrons to be on a stable configuration enough to be able to be shared. However, temperatures do need to be high enough for atoms to overcome activation energy barriers. In chemistry, an activation barrier is the energy difference between the reactants' initial state and the transition state during a chemical reaction. If you want a reaction to proceed, that minimum energy barrier needs to be overcome. Therefore, a planet, moon, or any other celestial body in which astrobiologists aspire to find life needs to be at a certain distance from its parent star so it experiences an optimal temperature that facilitates such chemical reactions.

Finally, the fourth element. Living systems contain self-replicating molecular systems that can support Darwinian evolution. Interestingly, while Darwin's theory of evolution emphasizes the occurrence and transmission of mutations in a random fashion from one generation to the next, living systems can generate complex molecules in a non-random manner. This distinct capability sets living systems apart from abiotic systems.

Researchers have come up with clever ideas to identify how complex a molecule is, inferring that such complexity can indicate the existence of life. The deoxyribonucleic acid molecule, or DNA, is an extremely complex molecule that carries genetic information in all living organisms. Complexity is, therefore, something that seems to be crucial for life. In that train of thought, a group of scientists who are part of NASA's Network for Life Detection (NfoLD) initiative are aiming to come up with a method to measure complexity.[12] These scientists have proposed using the *molecular assembly (MA) number* to experimentally determine the complexity of a molecule. In essence, the MA number defines the number of steps that are required to construct a molecule. Molecules with a high MA are very unlikely to have been produced by non-living organisms, or in other words, as the MA of a molecule increases, the more unlikely that such a molecule was formed by an abiotic process. This MA number could help us to detect the second type of biosignatures: the ones that we don't know of.

DNA is essential for Darwinian evolution. Darwinian evolution refers to the fact that living organisms on Earth have undergone a series of random structure variations. Variations that are advantageous to their survival pass through to or are inherited by the next generations. This passing of variations is done using the complex DNA molecule. Organisms that can transmit more advantageous variations for survival will prevail over others that failed to transmit essential variations. The descendants of organisms with advantageous mutations will be more likely to keep reproducing and transmitting those good traits, something that is known as *natural selection.*

False Positives

It is cumbersome to find methods to experimentally measure the characteristics of life to prevent as much as possible the occurrence of false positives. On the 7th of August 1996, the then-president of the United States, Bill Clinton, held a press conference at the White House. The topic: a meteorite, a rock from Mars, named ALH84001, which fell to Earth from space around 13,000 years ago and was found in the Allan Hills (AH) region in the Antarctic. But this was not just another rock from space. Scientists had published a paper[13] that stated that after carefully studying the meteorite, the presence of a carbonate mineral was found. A carbonate mineral is a mineral that contains the carbonate ion as a fundamental component in its chemical structure. The interesting thing is that they found that the carbonate mineral had formed small, spherical, or nearly spherical structures called globules. These *carbonate globules* were estimated by the researchers to have formed 3.6 billion years ago at high temperatures of around 973.15 Kelvin (700 degrees Celsius). Their investigation also concluded that the globules were indigenous to the meteorite, or in other words, they were formed while the rock was on Mars and not during the time between when it fell to Earth and when it was recovered.

This is not that hard to believe, as the remoteness and harsh conditions of the Antarctic make it a nearly perfect environment to keep any external factor from contaminating the recovered samples. The interesting thing here is that the texture and size of the globules have a remarkable resemblance to the carbonite globules found on Earth, which are the result of processes performed by bacteria. The researchers concluded that biotic processes can explain many of the observed features even though inorganic formation is possible. That's it, then? Astrobiologists found life on Mars, so what's the point of this book? Like many other times, the evidence was not conclusive. New research from 2022[14] revealed that geochemical processes, like water-rock interactions on Earth, could have created the features observed in the meteorite ALH84001. This is one of the discussions that the analysis of the samples from Mars collected by Perseverance will help settle in the near future.

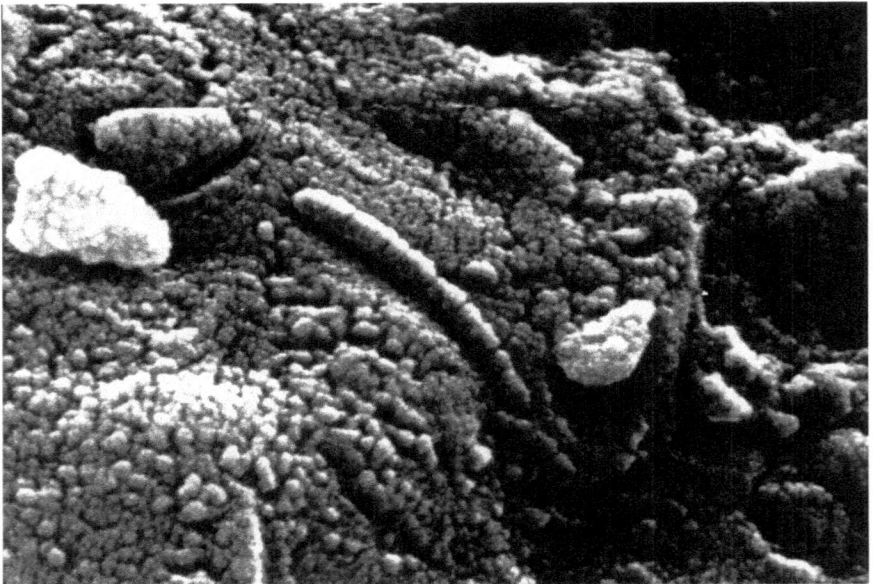

Fig. 6.9 The meteorite ALH84001. Does it contain microscopical and chemical evidence of life from Mars? Credit: Britannica.

Panspermia

An interesting question is raised. What happens if life is found on Mars and it resembles life on Earth? For instance, carbon-based life using DNA. If we completely rule out that our spacecraft, landers, or rovers were the ones bringing that life from Earth, then there is the possibility that life here on Earth originated on Mars and then transported to Earth somehow in the past. Note that the other way around is also a possibility, that life on Mars (if found) originated on Earth. This concept is known as *panspermia*. Panspermia is a scientific hypothesis suggesting that life, in the form of microorganisms, could have been distributed via meteoroids, asteroids, comets, or any other mechanism that allows material exchange between different bodies in the solar system (or between planetary systems). This means that we could all be Martians, or from any place where life originated first. Note that this does not actually answer the question of *how* life originated, only moves the origin to a different place.

Technosignatures

As many things in this field, we owe the first definition of technosignatures to Jill Tarter. In her article, *The evolution of life in the Universe: are we alone,*[15] Tarter defined a technosignature as the "evidence of some technology that modifies its environment in ways that are detectable". We humans, for instance, have been modifying our environment since the dawn of mankind. Consider deforestation, for instance. Once we settled and left our nomad way of life, we started cutting trees to make room for our cows, to plant, and to build our homes. This significantly modified the landscape of the surface of the Earth.

The search for technosignatures has been in the minds of many researchers and space enthusiasts for a long time. However, the field reached its deserved status when NASA established the Search for Extraterrestrial Intelligence (SETI) Program Office at Ames Research Center in 1976.[16]

Prior to that date, at a seminar planned by the astronomer Frank Drake in 1961, the first meeting on the topic of SETI occurred. The seminar took place at the Green Bank Observatory in West Virginia, and it is reported that at least a dozen people attended. Notable attendees included biochemist Melvin Calvin, who won the Nobel Prize in Chemistry that year, and a very young rising astronomer you may have heard of, Carl Sagan.

The seminar was part of project Ozma, considered the first modern attempt to detect interstellar transmissions. The search was run for six hours a day, from April to July 1960, using the receiver on the 26-meter antenna of the National Radio Astronomy Observatory (NRAO) at Green Bank.

The search was done by attempting to receive radio transmissions. Radio waves are commonly considered the most likely way of electromagnetic communication. This is due to their ability to be less affected by dust and other particles compared to shorter wavelengths. The longer wavelengths of radio waves make them less susceptible to scattering, which is the interaction with other particles in space. The reason is that dust particles are small enough to interact directly with wavelengths of similar sizes. Longer wavelengths suffer, therefore, minimal scattering.

Due to their long wavelengths, radio waves are also less affected by absorption. Absorption is the process by which electromagnetic waves are absorbed by the different elements in the medium, in this case, atoms, molecules, or dust particles.

Within the range of radio waves, the 1,420 MHz frequency is commonly the favorite one to listen to. This frequency corresponds to the 21-centimeter emission line of interstellar hydrogen, which is very relevant and pervasive in several astronomical events. Therefore, it is assumed to be a well-known frequency by any technologically mature society.

Another important characteristic of radio waves is that we humans are prolific at generating and using them. Our radio wave technology is very mature. We have perfected radio telescopes and our ability to detect such signals. This means that we currently have the observing capability to detect a possible technosignature if it is generated using this technology.

Axes of Merit

Observing capability is precisely one of the nine axes of merit for tech-nosignature searches that Dr. Sofia Z. Sheikh proposed on a remarkable paper[17] in 2019. Dr. Sheikh, currently a National Science Foundation – Advancing STEM Careers by Empowering Network Development (NSF-

ASCEND) Postdoctoral Fellow at the SETI institute, and who describes herself on her website as "a researcher who specializes in the Search for Extraterrestrial Intelligence", has proposed a framework for the search of technosignatures that would allow to comprehensively compare the different search techniques. Briefly, the nine axes of merit are as follows:

1. **Observing capability**. It refers to the technological ability of the whole field of astronomy at a given time to search for a given technosignature. This means that there are searches that can only be done in the far future, whereas there are others that can be done right now.

2. **Cost.** In this context, cost includes not only the financial aspect of it but also telescope time, computing time, staff resources, and any other associated cost.

3. **Ancillary benefits.** This refers to the fact that technosignature searches can also be used for other purposes. This is not only from a scientific point of view but also in contexts such as philosophy and education for instance.

4. **Detectability.** This axis deals with how strong a technosignature is so astronomers can detect it unequivocally.

5. **Duration.** This refers to the length of time that a technosignature would be detectable. If the civilization's lifespan is too short, we may never be able to detect one of their transmissions

6. **Ambiguity.** Technosignatures can be mistaken for natural phenomena. For example, unusual observed brightness patterns of a star may be confused with a megastructure around the star or simply explained by an uneven dust cloud moving around the star. This is what happened with the famous, or now infamous, WTF (yes, this acronym means what you think it means) star or Boyajian's star.[18] This star presents very unusual dimming in brightness which some have taken as evidence for an artificial megastructure around it. The true nature of such changes in brightness are not fully understood yet, but being part of a multi-stellar system, a circumstellar dust ring, or even a cloud of disintegrating comets are some of the possible explanations for such a behavior.

7. **Extrapolation.** Our confidence in detecting a technosignature will increase if we already possess and understand the underlying technology here on Earth. In other words, if we already have the technology, we may be able to detect it with more confidence. A famous quote by the notable sci-fi Arthur C. Clarke reads: "Any sufficiently advanced technology is indistinguishable from magic". For instance, a civilization that may have found a way to control and harness the power of black holes may be beyond our technological capabilities. This means that for many years astronomers will be unable to detect such a technology.

8. **Inevitability.** The search for technosignatures that are an inevitable use of technology according to our understanding of physics should be prioritized over those that rely on assumptions or unknown motivations. For instance, searching for waste heat similar to the produced by technologies we already have on Earth, should take precedence over looking for intentional communications, which imply social motivations that we exhibit here on Earth.

9. **Information.** The value of a technosignature would be proportional to the amount of data we could derive from it. If we find an artifact in the solar system left there by ET, the information that could be extracted from it would be monumental. However, if we indeed come across of the existence of a technology at a large distance, we would not be able to learn too much from it in subsequent follow-ups.

Dr. Sheikh's framework creates a very practical way of categorizing the different search techniques for technosignatures. Even more, she has made available the code used in the article to create the Figure 6.10,[19] allowing people to give value to the nine axes using their own technosignatures. I have created a graphical user interface wrapper around this code;[20] give it a go.

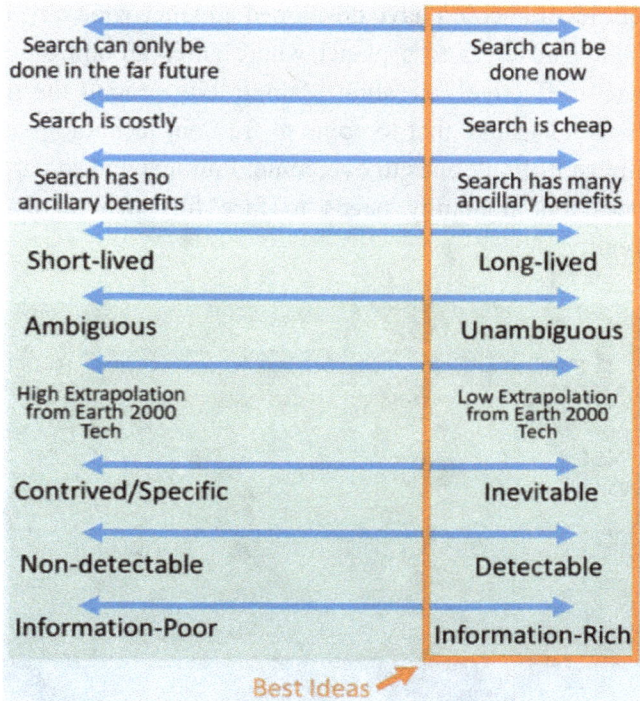

Search can only be done in the far future	Search can be done now
Search is costly	Search is cheap
Search has no ancillary benefits	Search has many ancillary benefits
Short-lived	Long-lived
Ambiguous	Unambiguous
High Extrapolation from Earth 2000 Tech	Low Extrapolation from Earth 2000 Tech
Contrived/Specific	Inevitable
Non-detectable	Detectable
Information-Poor	Information-Rich

Best Ideas

Fig. 6.10 Nine axes of merit for technosignatures search. Those ideas at the right hand of the framework are the ones that should be prioritized. Credit: Paper by Dr. Sheikh and modified in the report from the 2018 NASA Technosignatures Workshop.

For instance, with the invention of lightbulbs, our cities became a beacon of our presence during the nighttime. Anyone from out there with a potent enough telescope will be able to detect such features in our planet and would conclude that those are the result of intelligent life. However, current technology is still limited to detect such features on a potential habitable exoplanet. We simply don't have a telescope of a resolution that would allow us to detect features on the surface of an exoplanet, not even the one orbiting Proxima Centauri, the closest star system to Earth. Therefore, under the nine axes of merit framework, the detection of an extraterrestrial city is ranked with a low merit in the observing capability axis, but not for too long. Astronomers may be able to detect nightside city lights under certain circumstances. The brighter the city, the more chances astronomers get to be able to see it. It is expected that an advanced civilization could have developed a planet-wide city, or *ecume-nopolis*. This term refers to a planet where cities around the world are interconnected, effectively creating a single city around the planet. On Earth, for something like that to come to fruition, technology challenges are not the most difficult ones to overcome. Political and social issues are critical issues that humanity needs to face for such an endeavor to become a reality.

Fig. 6.11 A hypothetical planet where all the cities around the world have fused to create a world-wide city. Credit: The public Internet.

Professor Thomas G. Beatty, currently at the University of Wisconsin, has analyzed[21] how to detect potential cities on Earth-analog planets orbiting G, K, and M-dwarf stars. In addition, he has considered the detectability of city lights for exoplanets orbiting Sun-like stars within eight parsecs of the Sun. Finally, he has estimated the detectability of city lights on terrestrial planets orbiting nearby stars. As you have to start somewhere, in his calculations, Professor Beatty assumes that the lights from these cities are similar to those we use as street lights here on Earth. In his paper, Prof. Beatty suggested using two upcoming telescopes, NASA's Large UV/Optical/IR Surveyor (LUVOIR), and Habitable

Exoplanet Observatory (HabEx) to search for these potential exo-cities. LUVOIR, which is scheduled to be launched by mid-2030s, has as its main science goal, to study the atmospheres of identified rocky planets around various stars. LUVOIR will investigate planetary atmospheres for constituents such as water, oxygen, ozone, carbon dioxide, and methane, which are believed to be present in the biosphere of potential habitable worlds.

The HabEx mission is a conceptual mission that aims to directly image planetary systems around Sun-like stars. HabEx will also search for atmospheric water and gases that may indicate biological activity.

Prof. Beatty has concluded that using a 300-hour observation window, LUVOIR will be able to detect a planet-wide-city from a planet orbiting the star Epsilon Indi, a star located at approximately 12 light-years from Earth. More generally speaking, he has concluded that planet-wide cities for planets orbiting between 30 to 50 nearby stars will be detectable by employing both LUVOIR and HabEx. Therefore, we can conclude that the axis *Ancillary Benefits* in Dr's Sheikh's framework for this technosignature search will render many ancillary benefits and can be considered a good idea to pursue.

The Kardashev Scale

This idea of a planet-wide city is considered a characteristic of a presumed Type-1 civilization in the *Kardashev scale*. The Kardashev scale was proposed by the Soviet Astronomer Nikolai Kardashev in 1964. In his paper titled "Transmission of Information by Extraterrestrial Civilizations",[22] Kardashev classified hypothetical extraterrestrial societies in terms of their energy needs in a scale from 1 to 3. Kardashev's argument is centered on the assumption that, if a civilization wants to make its presence known throughout the universe, it must transmit a high-power signal that maintains the same intensity in every possible direction (isotropic). A signal with such features is necessary to overcome the incredibly long distances and background noise characteristic of deep space.

The key aspect here lies in the concept of "high-power". Generating a high-power signal requires high-energy sources. To do so, do you need to utilize the totality of the available energy in your home planet, star, or galaxy? To be able to harness Earth's complete energy potential, we would need to fully exploit our traditional fossil fuels (carbon, petroleum, coal, natural gas), and all the renewable energy we can get our hands on. In this context, the term renewable energy extends beyond the traditional and popular (or unpopular, depending on where you live and/or your political views) wind and solar natural sources of energy. 'Non-traditional' renewable energy sources encompass biomass energy, which is derived from plants and animal waste; ocean energy, which transforms the kinetic energy due to the movement of water into electricity; and geothermal energy, drawn from the Earth's internal heat, including that from volcanoes. My favorite non-traditional renewable energy source is tectonic energy, which derives power from the movement of tectonic plates (the mechanism on Earth that causes earthquakes).

So, how much energy are we talking about? Kardashev estimated that the total energy available on a planet will be in the order of a 4 followed by 12 zeros (4×10^{12}) or 40 terawatts (40 TW). To put this number in perspective, on average, traditional incandescent light bulbs use about 60 watts of electricity, whereas LED light bulbs use about 10 watts.

Type-I Civilization

Kardashev's estimation is based on the fact that all the radiation that reaches Earth from the Sun if converted into energy, would be in the order of 40 terawatts. The assumption is that a civilization that can harness all that energy belongs to the type-I category. The total Earth's electrical energy generation in 2022, including renewable and non-renewable sources, is estimated to be around 3.3 terawatts (3.3 TW).[23] According to Kardashev's original argument then, Earth's energy consumption is indeed of a one Type-I civilization already. However, as discussed before, in this exercise, solar energy needs to be considered, as well as all the different types of energies available to the inhabitants of a planet. To consider these other sources of energy, the Argentinian astronomer Guillermo A. Lemarchand redefined[24] the energy consumption of a Type-I civilization as one that generates 10 to the 16 power (10^{16}) Watts.

Fig. 6.12 An artistic representation of Earth as a Type-I civilization. Type-I civilizations can utilize all the available resources in their home planet. Credit: the public Internet.

So, is or isn't humankind a Type-I civilization? Carl Sagan had something to say about this. Sagan realized that having civilizations categorized with integer numbers (1-3) was not the best possible modeling of the evolution of civilizations. Therefore, he proposed a new classification schema, still based on Kardashev's types, but introducing a mathematical construct that allows the calculation of intermediate levels. Essentially, Sagan proposed a formula[25] that uses the power consumption of the civilization as input and spits out a number representing the Type of that civilization. Sagan calculated the current humanity scale to be 0.7 and defined humans as being at a "technological adolescence" age, with the ability to self-destruct. Using Sagan's formula, it can also be inferred that a civilization Type 0, which was not discussed by Kardashev, uses around 1 million Watts (1MW) of energy, which corresponds to the Stone Age in our case.

Regardless of humanity still not reaching a Type-I category, we have already developed something that could be considered a Type-I technology. In addition to the energy requirements already described for Type-I civilizations, the citizens of such civilizations are fully interconnected and united towards a single goal. Therefore, many authors deem the current Internet as a Type-1 technology in the sense that it allows humanity to be globally connected. This has become more evident with the latest advancements in smart cities, Internet of Things (IoT), and the emergence of digital currencies such as Bitcoin. We can already see that all these advancements in technology require more and more energy, hence the need for as much power as possible —despite the efforts of people like me that have devoted their entire PhDs to researching ideas on how to save energy.[26]

Type-II Civilization

When all the resources available on a planet are not enough anymore, a civilization must find alternate places to harness energy from. Kardashev hypothesized that a Type-II civilization would require, and therefore, be able to produce energy in the order of a 1 followed by 26 zeros watts (10^{26}). To achieve that, the civilization must develop a technology capable of capturing the majority or the totality of the energy produced by their home star; a megastructure around the parent star will do the trick. This is, of course, easier said than done. Such a hypothetical megastructure is known as a *Dyson sphere* and would provide a civilization with around 400 yottawatts (4 $x10^{26}$ watts) of power —essentially, the civilization's home star's entire energy output.

The science fiction writer Olaf Stapledon first proposed the concept of such a structure in his novel Star Maker in 1937. However, it was the physicist Freeman Dyson (1923-2020), who introduced the concept as a possible technosignature in his 1960 paper *Search for Artificial Stellar Sources of Infrared Radiation*.[27]

Similar to how we use solar panels here on Earth, a Dyson sphere would consist of orbiting satellites or arched panels covered with solar panels around the star. Those solar panels will irradiate infrared light that could be detectable using our telescopes. Therefore, detecting infrared excess around nearby stars could be an important step towards finding one of these Dyson spheres, and therefore, a Type-II civilization. Albeit such a discovery "would not by itself imply that extraterrestrial intelligence had been found", as warned by Dyson himself.

Fig. 6.13 Artistic impression of a Dyson sphere. Orbiting arched panels capture most of the energy of the star. Credit: Wikimedia Commons.

In the year 2000, researchers, including Kardashev from the Astro Space Center Lebedev Physical Institute from Moscow, Russia, proposed the term *Astroengineering Construction* (AC) as a more general term that includes Dyson spheres and other "building" in space around a central star. In their paper,[28] the authors explored the Infrared Astronomical Satellite (IRAS) database to search for evidence of Dyson spheres and other potential alien constructions. Astroengineering Constructions would absorb energy from different types of activities and re-emit it back to space as infrared radiation. The IRAS database was constructed as a result of a joint project of the US, UK, and the Netherlands. The IRAS survey includes 350,000 new infrared sources, and among other things, the survey helped to identify six new comets and revealed for the first time the core of our galaxy, the Milky Way. Nothing was found in Dyson spheres, but the researchers identified two temperature ranges of interest: 110-120 Kelvin (-163 to -153 degrees Celsius) and 280-290 Kelvin (7 to 17 degrees Celsius). Objects within these temperature ranges were found to be concentrated directly around the Galactic plane and Galactic center. These are targets that can be explored in further surveys at other wavelengths, including optical and radio.

A single solid structure around a star would be difficult to construct and maintain. Therefore, variants to the Dyson sphere have been proposed, being the *Dyson swarm,* one of the most popular. The Dyson swarm is comprised of living habitats, satellites, energy collectors, and self-replicating machines, which orbit around the star resembling a net. Energy is transferred wirelessly between each component and the home planet.[29]

Fig. 6.14 Many components, including self-replicating machines and orbiting solar energy collector circul the home star of a Type-II civilization. Credit: Wikimedia Commons.

Another variant of a Dyson sphere is a *Dyson ring*. A ring of solar power collectors is placed around the home star on a Dyson ring. This is the simplest and cheapest version of a Dyson sphere and one that a potentially advanced civilization could have used as their first step to collect energy from their parent star.

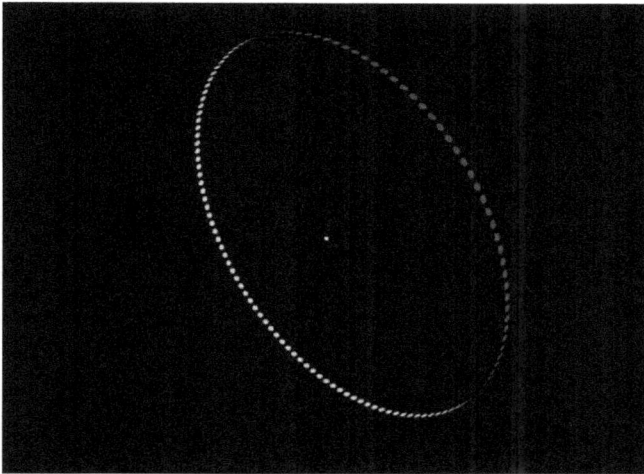

Fig. 6.15 A fleet of solar power collectors orbit the parent star forming a Dyson ring. Credit: Wikimedia Commons.

Type-III Civilization

If the energy produced by the parent star of a planet inhabited by a Type-II civilization is not enough, they will need to harness the energy produced by nearby stars. Eventually, energy requirements might take them to a point where the civilization would need to utilize the energy produced by all the stars, including black holes, neutron stars, and every conceivable celestial object in their home galaxy. The energy requirements are now estimated to be in the order of a 1 followed by 36 zeros (10^{36}) Watts. Researchers refer to such a hypothetical civilization as a Type-III. This implies, of course, that one of those civilizations has mastered inter-stellar travel and has, more than likely, colonized many of the worlds orbiting the stars in their galaxy.

A Type-III civilization is hard to conceive. In a very speculative way, the beings that comprise such a civilization would be some sort of immortal self-replicating AI machines that don't have the constraint of time or the ability to get bored in that sense, given the immense distances that inter-stellar travel requires.

Fig. 6.16 A Type-III civilization has managed to utilize all the resources within its home galaxy. Image generated by AI. Credit: Midjourney.

The same principles to detect a potential Type-II civilization also apply to detect a Type-III: infrared excess. In this case, astronomers direct their attention to entire galaxies instead of focusing on individual stars.

In 2015, Roger Griffith of Penn State University and collaborators compiled a catalog[30] of 93 candidate galaxies that exhibited unusually extreme infrared emissions. Prof Michael Garrett of the Netherlands Institute for Radio Astronomy (ASTRON) and Leiden Observatory capitalized on this catalog, conducting a search for potential artificial radio emissions emanating from the objects within the catalog that displayed the most distinct infrared signatures. However, he argued[31] that all these atypical infrared emissions could be explained through standard astrophysical interpretations. Consequently, Griffith concluded that Type-III civilizations are "either very rare or do not exist in the local universe".

Other Possible Technosignatures

Apart from Dyson Spheres, and extraterrestrial cities, researchers and SETI enthusiasts have proposed several possible technosignatures.

Technological Artifacts

A significant source of technosignatures could be the detection of technological artifacts. In all the missions that reached the Moon, starting with NASA's Apollo 11 in 1969, the crews performed several experiments (yes, despite all the crazy conspiracy theories out there, humans did land on the Moon in 1969) and left a number of technological artifacts behind. The list, which is not by any means comprehensive, includes seismometers, cameras, lunar module ascent stages, and Laser Ranging RetroReflectors (LRRR). These last ones still work and have been vital for measuring accurately the distance between the Moon and the Earth. Apollos 15, 16, and 17 left lunar rovers.

Fig. 6.17 The lunar module "Orion" and lunar roving vehicle photographed by astronaut Charles M.Duke during the Apollo 16 mission on April 21, 1972. The rover remains on the Moon. Credit: NASA.

However, the USA is not the only country that has left things behind on Earth's natural satellite. In 2023, the entire world celebrated the achievement of the Indian Space Research Organization (ISRO), India's national space agency. The ISRO launched the spacecraft Chandrayaan-3 aboard of the LVM3-M4 heavy-lift launch vehicle on July 14, 2023. The LVM3-M4 rocket departed from Satish Dhawan Space Centre's Second Launch Pad in Sriharikota, Andhra Pradesh, India. Chandrayaan-3 successfully entered the Moon's orbit on August 5, 2023; its payload included the lunar lander named Vikram, and a lunar rover named Pragyan.

On August 23rd, the lunar lander touched down near the lunar south pole. This magnificent achievement made India the fourth country until then (the USA, China and Russia the other three) in successfully landing on the Moon, and the first to do so near the lunar south pole.

Less than six months later, on January 19, 2024, the Japan Aerospace Exploration Agency (JAXA) successfully landed on the Moon, making Japan the 5th country to land a spacecraft. Japan's Smart Lander for Investigating Moon (SLIM), nicknamed *The Moon Sniper*, used a technology known as *pinpoint landing*. This technology allowed the spacecraft to land within a high-precision zone of just 100 meters wide. SLIM was carrying two small autonomous rovers — lunar excursion vehicles, LEV-1 and LEV-2, which were released just before landing.

These Indian and Japanese landers, rovers, and instruments will join their USA counterparts as technological artifacts left by humankind on the Moon.

As we previously discussed, several rovers, probes, and even a small helicopter (see Figure 6.18), among other things, have also been left behind on Mars. If a hypothetical alien crew mission in the future comes to perform some exploration in the solar system long after we are gone (we will discuss the concept of the 'Great Filter' in the next chapter), these artifacts will serve as evidence of our existence. It is logical to think that potential civilizations out there have also left behind scattered over their home planetary system, or even ours, evidence of their eagerness to explore.

Fig. 6.18 The Ingenuity helicopter at its final resting place, as imaged by NASA's Perseverance rover on Mars on February 25, 2024. The helicopter experienced an unrecoverable malfunction ending its mission. Credit: NASA/JPL-Caltech/LANL/CNES/IRAP/Paul Byrne.

Finding artifacts in the solar system possibly being left by another civilization is a worthwhile endeavor. Dr's Sheikh describes the search for technological artifacts as an enterprise that is something worth pursuing. In terms of the observing capability, we can search for them right now. The cost is low, as we can rely on existing instruments and resources. Ancillary benefits are also high as we could learn about the planetary surface processes in the potential body where such an artifact is found. The amount of information scientists could extract from such a finding is enormous. However, who knows if they would be able to decipher the language or understand the purpose of those artifacts.

Oumuamua

But how close are scientists to finding one of these artifacts? Well, for a very small number of astronomers, one being the eminent Avil Loev the most vocal —as stated in his book *Extraterrestrial,*[32] this has already

247

happened. In 2017, an interstellar visitor entered and then exited the solar system. The object, named *Oumuamua*, which means 'a messenger from afar arriving first' in Hawaiian, was the first interstellar body detected passing through the solar system. The object was observed using the PAN-STARRS telescope at Haleakala Observatory, in Hawaii, on October 19, 2017. Unfortunately, when it was first observed, Oumuamua was already heading away from the Sun and very far away from Earth.

Oumuamua is part of a selected group of objects known by *interstellar objects* (ISO). Contrary to what we have learned in Hollywood movies, this term does not mean that such objects are of an artificial extraterrestrial nature, but that they came from a different planetary system. Such a conclusion can be reached due to the angle of obliquity at which the object travels. Let's remember from our discussions in Chapter 1 and 2, that the most accepted planet formation hypothesis is that all the elements in a planetary system come from the same molecular cloud, and, therefore, they are located within the same plane. Astronomers can identify that an object is not indigenous to a given planetary system if it is not within the same plane as the rest of the objects in that system. In addition, Oumuamua's high speed revealed that it was not bound to the Sun's gravity. When it passed through the solar system, it exhibited a remarkably high speed, which means that it had enough energy to escape the Sun's gravitational pull, meaning that there was another star to which it was bound. Being an interstellar object by itself is an odd characteristic, as this was the first time something like that has been detected. But the odd things did not stop there. For instance, when Oumuamua was close to the Sun, no evidence of a tail or a coma —what surrounds a comet, was observed. It also seemed to have accelerated as it left the solar system. From its large variations in brightness and color, it was estimated that the interstellar object measures between 100 to 400 meters and about ten times as long as it is wide, or in other words, it has a cigar shape. Due to these weird characteristics, one possible explanation was that Oumuamua was an alien probe sent to explore the Sun and the rest of the solar system.

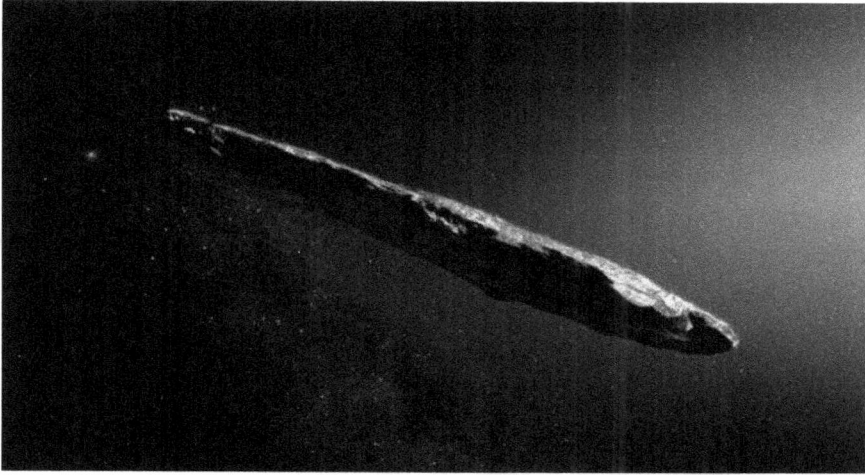

Fig. 6.19 Artistic depiction of Oumuamua. This cigar-shaped rock is the first interstellar object detected. Credit: European Southern Observatory / M. Kornmesser.

However, astronomers have provided several scientific arguments to prove that Oumuamua is not an extraterrestrial spacecraft. For instance, Dr. Jennifer Bergner, a UC Berkeley assistant professor of chemistry, explains that the acceleration observed as Oumuamua left the solar system is consistent with hydrogen gas release due to hydrogen sublimation. Contrary to regular comets, which eject carbon monoxide or carbon dioxide, a comet ejecting hydrogen would lack a coma or tail as hydrogen is less massive and does not have the momentum to pull too much dust.

The odd shape can also be explained as a result of extensive tidal fragmentation. The idea is that Oumuamua was once part of a parent body, in this case, a hydrogen iceberg,[33] that had a close encounter with the parent star. This encounter caused the iceberg to be fragmented, being Oumuamua one of the resulting fragments. The controversy continues. But I can't help myself and need to bring back Dr. Sagan's quote one more time: "Extraordinary claims require extraordinary evidence." A quote that, surprisingly, Dr. Loeb seems to dislike, as expressed in his book *Extraterrestrial*.

The natural nature of Oumuamua seems to have been reinforced with the discovery of a second interstellar object: the comet Borisov. This comet, which exhibits evidently cometary appearance, was discovered by Gennady Borisov in August 2019. The orbit of this object has been found to be highly hyperbolic with respect to the Sun —contrary to elliptical or circular— with speeds of approximately 32 kilometers per second. The hyperbolic nature of the orbit indicates that the object was able to enter the solar system, pass the Sun, and then continue its trip into outer space. Such characteristics are consistent with the orbital characteristics of Oumuamua.[34]

Fig. 6.20 Comet Borisov as imaged by Hubble in October 2019 when it was at about 418 million kilometers from Earth. Credit: NASA, ESA and D. Jewitt (UCLA).

Industrial Waste

Industrial waste or pollution has also been proposed as a detectable tech-nosignature. Waste gas products from technologically advanced extrater-restrial civilizations may be present in the atmosphere of an exoplanet to the point that could be detectable. This is what Dr. Sara Seager from the Massachusetts Institute of Technology (MIT), and a group of collabora-tors propose in their research.[35] They argue that life avoids producing or using certain gases based on fluorine, and those are specific waste of industrial production. Scientists, therefore, will be able to look in the spectra collected by observatories such as the JWST to detect these gases. The key here is how long astrobiologists need to observe a planet to detect these industrial pollutants. Known gases, such as chlorofluorocarbons (CFCs) or hydrofluorocarbons (HFCs), require large observation times, which is difficult to obtain given the limited amount of space observatories and the large number of observation proposals. On the contrary, the gases proposed to be detected by Dr. Seager and her team are non-carbon, fully fluorinated compounds that exhibit low water-solubility, which means that they will not dissolve in rainwater and fall to the ground or the sea. This increases the probability of having a large lifetime and concentrations of such compounds in a planet's atmosphere, which would result in shorter observation times required to detect them.

Fig. 6.21 Left: Pollution can have a temporary and detectable effect on the atmosphere of a technological advanced extraterrestrial civilization's host planet. Right: image of Earth's ozone layer hole (in blue) over Antarctica in 2019. Since the banning of CFCs, the ozone layer has recovered substantially. Credit: NASA.

Have We Already Found Them?

Many technosignatures and many ideas. However, so far, scientists have not had any luck. Astronomers have been searching for more than 60 years and have not found, beyond any reasonable doubt, any proof that extraterrestrial beings are indeed out there.

But the universe is 13.8 billion years old and incredibly vast. This means that although 60 years seem like a long time from a human lifespan perspective, in the grand scheme of things, it is quite a short period.

But how much of the universe has been searched? Penn State Astronomy Professor Jason T. Wright and collaborators have concluded on a 2020 paper,[36] that the fraction of the universe that has been searched to date is roughly "similar to the ratio of the volume of a large hot tub or small swimming pool to that of Earth's oceans". One can argue that concluding that extraterrestrials do not exist because any signal has been detected in 60 years would be the equivalent of taking a large hot tub to the ocean, filling it with water right there at the beach, and then concluding that sharks do not exist because none were found within that collected volume of water.

The Wow! Signal

After all these years, the closest astronomers have been to finding a non-human-originated technosignature is the famous Wow! signal. The signal is so important and intriguing that a myriad of books and articles have been written about it, being the "The Elusive WOW – Searching for Extraterrestrial Intelligence" by Roberth H. Gray,[37] in my opinion, probably one of the most comprehensive you can find.

On Monday, August 15, 1977, Dr. Jerry R. Ehman was a volunteer researcher working with the Big Ear radio telescope from Ohio State University. Big Ear had become the first full-time search for extraterrestrial life facility back in 1971. On that Monday, just after 11:15 am, the system registered a signal that lasted for about 72 seconds and was printed on paper by the computer with the rest of the results from the observations. For the untrained eye, the printing is just a bunch of numbers and letters that make no sense. However, Dr. Ehman noticed something in the printout that was surprising and exciting. It was exactly what he and other SETI researchers have been waiting for (and continue to wait for): a radio signal from a celestial source that is much stronger than the background noise. When he saw the printing and its potential, it wrote Wow! in the piece of paper.

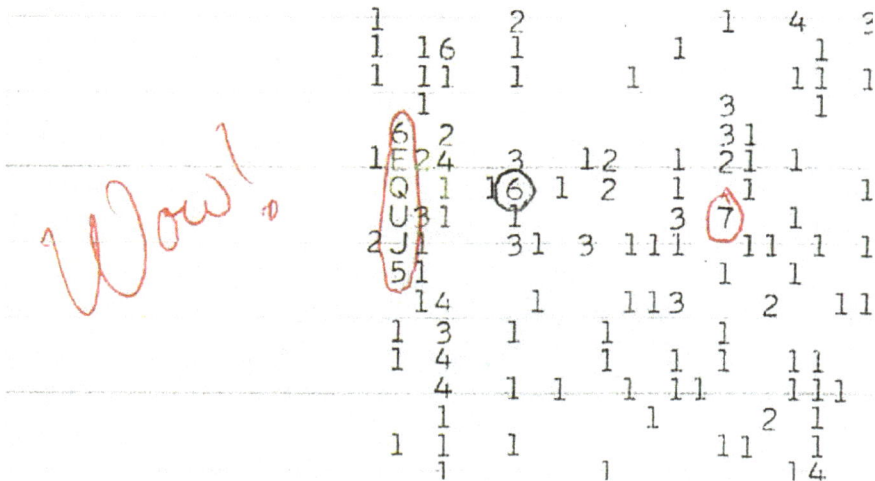

Fig. 6.22 A copy of the original printout that shows Dr. Ehman's "Wow" annotation. Credit: the public Internet.

Unfortunately, the signal was never observed again. But the people involved in the analysis discarded its origins from a natural radio source, given that no natural sources were found in the vicinity of that part of the sky where the signal was detected. In addition, no records of spacecraft were found during the time of the detection.

Like the signals we generate here on Earth, potential extraterrestrial signals are expected to be narrowband. This means that the signal only covers a small fraction of the frequency spectrum. Broadband signals (the opposite to narrowband) which are characteristic of most natural sources, are noise-like and spread over a wide band of frequencies. Broadband signals require large amounts of power if they are created artificially.

On the contrary, narrowband signals require less power and are ideal for long-range communications. In a narrowband signal, the information is transmitted on a fixed frequency, which is ideal for the receiver instrument as it can be tuned to listen on a specific range.[38]

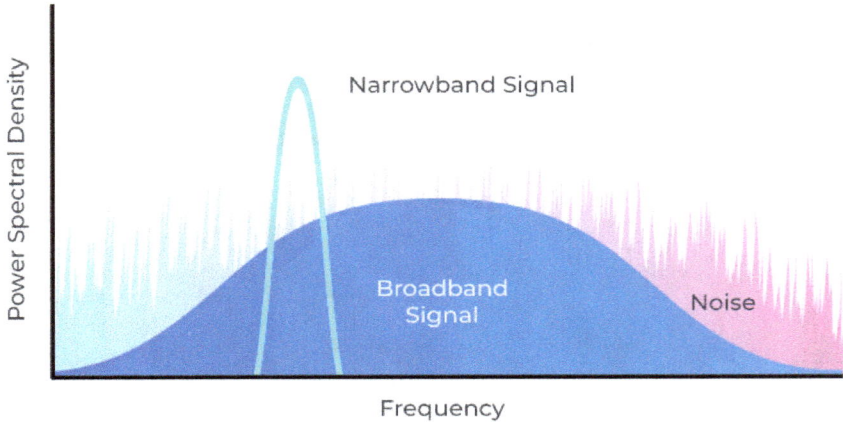

Fig. 6.23 Broadband signals spread across a range of frequencies whereas, a narrowband signal is focused on a fixed frequency. Credit: Sparkfun.

Narrowband signals, therefore, are easy to distinguish from the signals generated by most natural astronomical events. Well, the Wow! signal was indeed a narrowband signal centered in the 1,420 MHz frequency. Looks familiar? Yes, this is the frequency we mentioned earlier, which SETI researchers expect a technologically advanced civilization may use for transmission due to its significance in the universe.

Being narrowband by itself and centered on that particular frequency are not the only things that make the Wow! signal a strong extraterrestrial radio signal candidate. A pattern was also recognized accompanying the main central frequency. Similar to what happens in the book and movie of the same name, "Contact", a potential extraterrestrial broadcaster might choose to transmit a pattern, such as the series of prime numbers 2, 3, 5, 7, 11, 13, 17..., to prove that their signal is artificial. The pattern that was encoded in the Wow! signal looks extremely similar to the Lyman series of spectral lines emitted by hot hydrogen at ultraviolet wavelengths. Such a pattern is something that is well known to astronomers and physicists and, therefore, a fact that contributes to the importance and mystery of the signal.

However, two researchers, astronomer Antonio Paris and Evan Davies from St. Petersburg College in Florida, argue in a 2015 paper that, contrary to what SETI enthusiasts believe, the Wow! signal could have been originated because of passing comets.[39] Specifically, their paper highlights that the comets 266P/Christensen and P/2008 Y2 were transiting in the vicinity at the time when the Wow! Signal was detected.

Nuclei of comets are surrounded by large hydrogen clouds, with extensions of several million kilometers in radius. The hydrogen clouds would explain the main frequency of the transmission.

Fig. 6.24 Could comets be the source of the mysterious Wow! signal?
Credit: the public Internet.

Regardless of this strong hypothesis, many unknowns remain and the Wow! signal still prevails as the most promising candidate of a detected extraterrestrial communication.

A Story of More Than 5000 Worlds

Przybylski's star

What if a technologically advanced extraterrestrial civilization is capable (and keen) of using their home star to dispose of their nuclear waste? This is something researchers Daniel Whitmire, and David Wright suggested in an article[40] that appeared in 1980 in the Icarus planetology magazine, then directed by Carl Sagan. The extraterrestrial civilization could use such a capability to keep their home planet free from such wastes or as a way of advertising their presence to the rest of the universe.

Colloquially known as "star salting", this hypothetical practice has been suggested as a strong technosignature. According to the current under-standing of stellar processes, if this is happening somewhere in the universe for a given star, astronomers would see very distinctive traces of elements that should not be present in the spectra of the star.

What do I mean by "elements that should not be present"? As discussed in Chapter 1, stars generate their energy through nuclear fusion, starting with the fusion of hydrogen into helium and progressing through to heavier elements like iron. Thus, it is common to find a mixture of elements, including iron, in most stars. However, heavy elements such as promethium, with a half-life of less than 20 years, and plutonium, with a half-life of up to 24,000 years, are not expected to be found in large quan-tities in a star. These elements, if present, should be in small portions as they would have decayed long ago, given the extensive ages of stars. A significant presence of such elements in a star might indicate chemical or nuclear processes that are still unknown or a replenishing mechanism that continues to add those elements to the star. In summary, a typical star should contain large quantities of iron; heavier elements than iron are expected to be found only in very small quantities.

Well, brace for it. There is a star that exhibits quite the opposite of what is expected: very low traces of iron and large quantities of heavy elements. The star HD 101065, better known as *the Przybylski star*, was discovered by the Polish-Australian astronomer Antoni Przybylski (1913-1985) in 1960.[41] Its peculiar spectrum continues to challenge our understanding of stellar physics and has many researchers scratching their heads.

The Przybylski's star, located at approximately 310 light-years from Earth, is categorized as an Ap-type star; the 'p' in the name of this category stands for "peculiar". Ap-type stars rotate slower than regular type-A stars, favoring the reliable chemical composition measurement.

Fig. 6.25 Is an extraterrestrial civilization dumping their nuclear waste onto their home star? Credit: image generated by OpenAI's DALL·E.

Multiple hypotheses have been proposed to explain the unusual spectrum of the Przybylski's star. Following the concept of *Occam's razor*, which suggests that the simplest explanation is most likely correct (and also the most boring in this case), the most probable explanation is that the measurements are simply wrong, and somehow, the presence of these heavy elements is due to some misinterpretation of the data. However, a few other more exciting explanations have been proposed. One of them is the proposition of an "island of stability". This idea suggests that super heavy elements, still yet to be observed in nature, exist, and could decay into the elements astronomers observe in the spectrum of the Przybylski's star. Another hypothesis suggests that a neutron star companion is "contaminating" the Przybylski's star's atmosphere. Due to the immense pressures and temperatures present in a neutron star, elements heavier than iron are likely to be present. Unfortunately, nothing seems to indicate the presence of such a companion object. That leaves us with the possible explanation that an extraterrestrial civilization is involved. As suggested by Whitmire and Wright in their 1980 article, an extraterrestrial civilization could be using "their local star as a repository for radioactive fissile waste material." If that is the case, such a practice would cause changes in the stellar spectrum of the star over long periods of time. Something that everyone agrees on is that more data and more studies are needed. The understanding of what is happening with this star can have a profound impact on our understanding of stellar and chemical processes, and potentially on the life as we know it as well.

Using Artificial Intelligence to Find E.T.

But the search continues, and with the advancement of technologies such as Artificial Intelligence (AI), Machine Learning (ML), and disciplines like data science, the SETI field has experienced a new air. The whole field of astronomy, to be precise.

Astronomy benefits greatly from the progress of data science. Among all the sciences, one could argue that the field generates the largest volume of data. An example is the Square Kilometer Array Observatory (SKAO), an intergovernmental organization that consists of two telescopes in South Africa and Australia and has its global headquarters in the UK. The telescope array is still being completed, and it is estimated that, when in full operation, SKAO will require the transmission of an average of 8 terabits per second for the facility in Australia and 20 terabits per second for the facility in South Africa. In total, estimations indicate that the SKAO will archive 300 petabytes of data per year. The access, storage, and analysis of all that information falls under what is known in the computing industry as *big data*. Big data refers to data that is so large, fast, or complex that it's difficult to process using traditional methods.

Humans are good at a number of things. We are creative, spontaneous, and intuitive. Computers don't have these capabilities (yet), but something that computers do well and a lot better than humans is to look for patterns. Data science, and more specifically, applied machine learning, is the future of SETI , astronomy, and science in general.

Machine learning algorithms fall into two categories:

1. **Supervised learning**: where classes of objects are predetermined, and suitable sets of training, validation, and testing examples are provided to define the optimal classification of objects and
2. **Unsupervised learning**: where the data themselves can determine the numbers of the different classes of the objects analyzed.

Given that the second type of learning requires little or no human intervention, it is considered a least-biased approach. *Deep-learning* is an unsupervised learning technique that trains the computer to learn on its own by recognizing patterns. Deep-learning is the technique responsible for the significant advancements in speech recognition, image identification, and predictive modeling. These technologies are becoming increasingly prevalent in our society, including within our homes and in our daily lives, thanks to the widespread adoption of digital assistants and conversational agents such as ChatGPT, Copilot, and Gemini.

Astronomers are taking advantage of these advancements in artificial intelligence and machine learning. A group of researchers in January 2023 published[42] the result of a deep-learning search for technosignatures of 820 nearby stars. This project analyzed 480 hours of on-sky data collected from observations done with the Robert C. Byrd Green Bank

Telescope in the US. Researchers applied machine learning techniques in the frequency range between 1.1 to 1.9 Gigahertz (GHz) to identify any potential extraterrestrial intelligence (ETI) signal. To train the model with "true" data, scientists generated simulated events by artificially injecting ETI signals into a subset of the collected data. The initial results returned a total of nearly 3 million signals of interest.

Eight promising signals were identified after a more careful analysis, which discounted radio frequency interference from human sources, such as GPS signals. These eight candidates come from five different stars within 90 light-years from Earth. Unfortunately, re-observations of these target stars did not reproduce the results obtained, which means that regardless of the nature of the signals, they are not persistent in time, or in other words, they are *transient events*. Despite these results, those promising targets can be revisited in future surveys.

To recognize a signal that could potentially be extraterrestrial, it is paramount to remove any potential human-made signals, referred to in this context as false positives. Even regular electronic appliances such as an innocent microwave oven can contaminate the collected data despite most observatories being in radio-quiet zones. This happened once at the Parkes radio telescope in Australia as reported in a paper in 2015.[43] Machine learning algorithms, therefore, are trained to recognize such patterns and remove them from the analyzed data.

In another interesting project,[44] researchers also used data-driven machine learning techniques to find the landing module of the Apollo 15 on the surface of the Moon. The study is a proof of concept that demonstrates that such a technique can be applied to other objects of the solar system and the Moon itself to potentially find technological artifacts left behind by a hypothetical extraterrestrial civilization. The Moon was used as a test bed for this proof of concept due to the wealth of satellite data available and the fact that numerous exploration missions have left non-natural relics on the surface. With the aim of training the algorithms, the researchers used images from parts of the Moon's surface where no human-made artifacts were present. The algorithm was then told that such images were *non-anomalous* (normal) samples. On the contrary, parts of the Moon's surface with human-made artifacts, such as the lunar lander modules left from the Apollo 15 and 17 missions, were categorized as *anomalous* samples. Three experiments were conducted, in which the algorithm was fed with a large number of images, the majority of which did not contain any artifacts. For instance, one of the experiments utilized a dataset of 8000 images around the Apollo 15 mission landing site, including ten images displaying artifacts left by the mission's crew. At the end of the exercise, the computer, trained using unsupervised learning techniques, successfully identified the landing module of the Apollo 15 mission. Works like these serve as excellent demonstrations of the potential of artificial intelligence and how astronomers can employ it in the search for this type of technosignatures elsewhere in the solar system.

Messaging Extraterrestrial Intelligence (METI)

So far, all the efforts we have discussed can be described as reactive endeavors. Astronomers listen or search for technosignatures hoping that some civilization out there has taken the first step in interstellar communication. However, initiatives such as Messaging Extraterrestrial Intelligence (METI), intend to take a more proactive approach. Some of the proponents of METI initiatives propose to create a message and send it to the stars. The idea is to let anything out there listening know that "you are not alone".

Technical aspects of such an enterprise will require, among other things, the definition of at what wavelength we should transmit on; how much energy is required; what are the targets where the message will be transmitted; and what would be the structure of the message. However, more philosophical aspects relate to questions such as why we should transmit an interstellar message? Or what are the dangers of pursuing METI?

It is worth noting that humans have already sent messages to the stars. The first and most well-known message is the *Arecibo message* sent from the Arecibo radio telescope in Puerto Rico in 1974. The target of the message was the globular star cluster M13, which is approximately located at 25,000 light-years from Earth. Some of the people involved were SETI's pioneer Frank Drake, Carl Sagan, among others. The message transmitted in binary, zeroes, and ones, contained the numbers one through ten, the atomic element of five elements that make up our DNA, a graphic figure of a human, the human population of Earth at the time, a graphic of the solar system, and the physical characteristics of the antenna dish used to transmit the message.

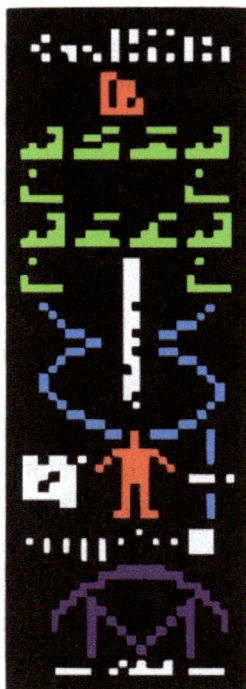

Fig. 6.26 Graphical representation of the Arecibo message. The message included an image of the Arecibo telescope, the solar system, DNA, a figure of human and some details about the biochemicals of earthly life. Credit: SETI institute.

Perhaps the second most well-known interstellar messages humans have sent are the ones onboard the space vehicles Pioneer 10 and 11, and Voyager 1 and 2.

Pioneers 10 and 11 were the first spacecraft to pass through the asteroid belt. Pioneer 10, launched in March 1972, was the first spacecraft to make direct observations and close-up images of Jupiter and the first human-made object to pass the orbit of Pluto. After its primary mission, it continued to explore the solar system's outer regions until March 1997. Its sister spacecraft Pioneer 11 was the first human-made object to fly past Saturn and to return the first images of the polar regions of Jupiter. The last transmission of the spacecraft was received by NASA in September 1995.

The Pioneers carried gold-anodized aluminum plaques on board. The plaques show an image of a human male and female along with symbols designed to serve as instructions for a possible extraterrestrial intelligent life to help them find Earth.

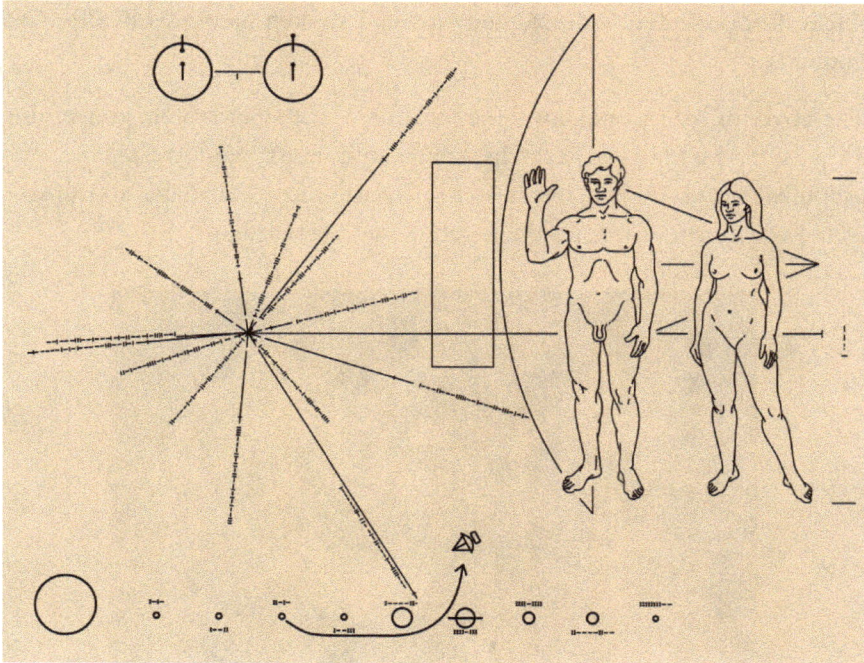

Fig. 6.27 Plaque on board of the Pioneer 10 and 11. The plaques contain the position of the Sun relative to 14 pulsars and the center of the milky way, along with a representation of the planets of the solar system. Credit: NASA.

The Voyagers, launched in 1977, had the primary mission of exploring Jupiter and Saturn. Once the primary mission was completed, these probes continued their space journey. These instruments have become the first human-made interstellar objects. In August 2012, Voyager 1 officially entered interstellar space, while Voyager 2 achieved this milestone in November 2018. The probes are expected to continue sending data until at least 2025. Until that date, the onboard science instruments are expected to continue to operate with the remaining electrical power and fuel. One characteristic that captivated the public's imagination and attention was the *Golden Record*. The Golden Record is a 12-inch gold-plated copper disk that contains sounds and 115 images from Earth. The record includes natural sounds such as those made by wind, thunder, birds, whales, and other animals. They also contain a selection of songs and music from different cultures and eras and spoken greetings in fifty-five languages.

The cover of the record shows the location of the solar system concerning 14 pulsars, the drawing of the hydrogen atom in its two lowest states, and graphical instructions about how to play the record, and the appearance of the waveform of video signals found on the recording.

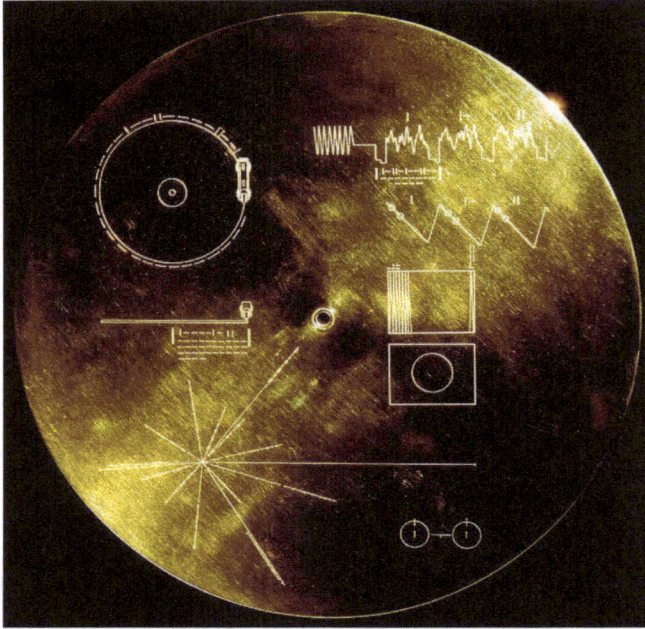

Fig. 6.28 A message to the stars. This golden record, which contains, among other things, instructions on how to find Earth, has left the solar system. Credit: NASA.

Detractors of sending messages, such as the ones sent by the Arecibo telescope and the ones on board the Pioneer and Voyager probes, often cite examples of what has occurred here on Earth when an advanced civilization has encountered a less advanced one. They fear that humans could face a situation similar to the one experienced by the indigenous people in South and Central America when they encountered the European conquerors in the fifteenth century, just to mention one example. However, arguments such as the espoused do not fully capture the whole spectrum of possibilities of making contact with an extraterrestrial civilization. The possible benefits of communication and cooperation with one of those societies may be worth the risk. However, for such a cooperation to happen, we need to find these extraterrestrial civilizations first. What if they are using technologies that we just can't comprehend, are impractical for us, or we don't even know exist?

SETI and the Multi-Messenger Astronomy

Several SETI researchers and sci-fi writers have proposed the possibility of extraterrestrial civilizations using gravitational waves or neutrinos as a form of communication.[45]

Gravitational waves, as we explored in Chapter 1, are ripples in the very fabric of spacetime caused by extremely violent and energetic processes. We discussed how the merging of neutron stars, or black holes, are mainly the events that astronomers know are responsible for such waves. A Kardashev Type II or Type-III civilization theoretically can manipulate these types of celestial objects. Regardless of their potential ability to perform such tasks, employing these significant energy intensive events to communicate seems quite unnecessary, especially, for day-to-day communications. Perhaps they could use this type of technology to send a once-in-a-lifetime message to advertise their presence to the rest of the universe.

On the other hand, neutrinos have been proposed as an alternative to electromagnetic waves for communication.[46] Neutrinos are fundamental particles ejected from a variety of subatomic reactions, like those occurring in the interior of stars. These ghostlike particles have extremely low mass and interact very weakly with matter. Not interacting with matter makes them ideal as a form of communication in environments where electromagnetic waves can be easily absorbed, scattered, or reflected, or for periods in which the communications link is blocked. However, employing neutrinos as a telecommunications system is extremely impractical for humans at the moment. For instance, the IceCube Neutrino Observatory is a cubic-kilometer-sized ice telescope designed to observe neutrinos. The instrument contains enough mass that one of every million neutrinos going through IceCube will eventually hit something, which results in a flash of light that can be detected and measured. In June 2023, the IceCube team made the announcement of the mapping of the Milky Way using neutrinos.[47]

Fig. 6.29 The IceCube observatory located at the South Pole. Credit: Josh Veitch-Michaelis, IceCube/NSF.

The Milky Way has long been suspected to be a source of high-energy neutrinos, and the IceCube team has now confirmed this. Additionally, the team used the detected neutrinos to produce an image of the galaxy.

Fig. 6.30 The first map of the Milky Way in Neutrinos. Credit: IceCube collaboration.

Neutrinos, therefore, represent another resource added to the multi-messenger astronomy toolkit. However, much like gravitational waves, the resources and energy required to build and operate a communication system based on neutrinos seem far-fetched. Regardless of how far-fetched such communication systems look to us, for other civilizations, it could be business as usual.

But let's forget for a moment whether we could detect or communicate with a technologically advanced extraterrestrial society. Could these extraterrestrial societies be real? Not everyone is so optimistic, and many arguments have been given over the years about why they might not exist. We will explore some of these arguments, collectively known as the solutions to the Fermi Paradox in the final chapter.

Chapter 7
The Great Silence – Are we alone?

TL;DR

One of the main points of the detractors of the Search for Extraterrestrial Intelligence (SETI) initiatives, is that astronomers have been looking for signals for more than 60 years, and nothing has been found so far. However, this argument doesn't consider the vastness of the universe and how little it has been searched. Regardless, if life is common out there, we should have heard from someone already. Humans went from not being capable of flying to becoming explorers of the solar system in less than 100 years. Considering how old the universe is, and some estimations that some Earth-like planets could be at least 2 billion years older than Earth, we should have heard of any hypothetical civilization that has the means of achieving interstellar travel.

In the summer of 1950, the Italian American physicist Enrico Fermi asked some of his colleagues during a lunch break: "Where is everybody?", referring to the lack of evidence of extraterrestrial civilizations. This question is popularly referred to as *The Fermi Paradox*. Since the question was formulated and popularized, people from different backgrounds have tried to answer it. There are thousands of books and articles dedicated to the subject. Possible answers to why we have not heard of anyone are known as the solutions to the Fermi Paradox. Traditionally, a good portion of the solutions are framed within the context of the *Drake Equation*, a probabilistic argument introduced by Frank Drake in 1961. Drake's goal was to stimulate a scientific dialogue around the existence of technological civilizations. The Drake Equation aims to estimate the number of technologically advanced civilizations in the Milky Way. Solutions considering the terms in the Drake Equation discuss the possibility of the universe not having enough planets or planets not being at the right distance of their home star to allow life to flourish. Other solutions bring the topic of how long a civilization could survive before being able to make their presence known in the universe. Such civilizations could have gone extinct before their signals were able to reach other stars, succumbing to some sort of universal great filter.

The *Great Filter* is the idea that human-like intelligence is required to go through multiple hurdles before they can colonize the universe. Other civilizations might just be silent on purpose, afraid of any unknown danger out there.

The Drake equation has been criticized and loved by many since its inception, and new versions have been proposed. One of the most popular versions is undoubtedly *The Seager Equation*, presented by MIT Professor and astrobiologist Sara Seager. Professor Seager proposes a probabilistic argument that is based on our current tools and limitations and calculates the odds of finding life based on the detection of biosignatures.

Fig. 7.1 Extraterrestrial civilizations should be common, but we don't hear from anyone. So, where is everyone? Credit: image generated by OpenAI's ChatGPT.

Search for Extraterrestrial Intelligence (SETI) Initiatives

Even with the fast pace of life, people continue to look up to the night sky and wonder if someone, like us, is on a distant planet, wondering the same.

Mark Twain's famous quote reads: "Find a job you enjoy doing, and you will never have to work a day in your life". I have always been jealous of two types of professionals: football players and scientists, who, to me, have that type of job. Astronomers, and those searching for extraterrestrial intelligence (SETI) in particular, get paid to collect data, analyze that data, and write about their findings (leaving the admin tasks and the politics aside, which unfortunately, consume much of their time). That, to me, fits perfectly with Mark Twain's quote. Unfortunately, not everyone shares the same enthusiasm about finding life elsewhere. Such endeavors have been categorized as futile, time-wasting, and ridiculed by many people, USA politicians in particular.

In February 1994, John Gibbons, President Clinton's science advisor, made it clear that he totally misunderstood the nature of SETI. He opined: "we've done a lot of observing and listening already, and if there were anything obviously out there, I think we would have gotten some signal".[1] Gibbons' comments were made after the US Congress had already canceled funding for NASA's SETI initiatives[2] in 1993, after years of SETI members practically begging for financial support.

The lack of funding forced the SETI institute to be established as a private, non-profit institution on November 20, 1984, by the CEO Thomas Pierson (1950-2014) and SETI scientist Jill Tarter. A funding model that persists until this day. The SETI institute relies heavily on donations to continue its work. SETI initiatives have struggled to get the traction that they deserve. Fortunately, this is changing.

On July 20, 2015, Stephen Hawking, Yuri Milner, and Lord Martin Rees, announced a new SETI initiative named Breakthrough Listen. This initiative was the first of several privately funded SETI endeavors. The Breakthrough Listen research program[3] aims to listen for messages from the 100 closest galaxies to Earth. The search, which is done with instruments that are up to 50 times more sensitive than existing telescopes, covers more than 10 times of the sky than previous programs and five times more of the radio spectrum. In addition, and compared to traditional searches done in the past, the program is performing the "deepest and broadest ever search for optical laser transmissions".[4] The most important aspect of this is the financial funding which has been secured for ten years for a total of 100 million US dollars.

As part of the Breakthrough Listen search for intelligent life, scientists have observed thousands of nearby stars in the Milky Way Galactic Center on several frequencies. Unfortunately, so far, we haven't received any positive confirmation of the existence of life elsewhere. So, one more time, where is everyone?

The Fermi Paradox

Leaving aside the argument we discussed in the last chapter about scientists having covered only a small fraction of the universe in their searches, if life —intelligent life in particular— were to flourish everywhere, we may have heard from someone by now. After all, the universe is billions of years old, and given that we went from being a society that did not have the capacity to fly to one capable of setting foot on the Moon in less than 70 years, not hearing from someone seems contradictory.

Such a contradiction has a name: The Fermi Paradox. History indicates that the Italian American physicist Enrico Fermi, who has been called the "architect of the nuclear age" due to his contributions to the Manhattan Project, was having lunch one day in the summer of 1950. While enjoying his meal, Fermi engaged in a casual conversation with fellow physicists Edward Teller, Herbert York, and Emil Konopinski about a cartoon by Alan Dunn in the *New Yorker* that depicted aliens stealing trash cans from the New York streets. The cartoon was referring to two recent stories: one about public trash cans that had been disappearing from the streets of New York and another that discussed the increased reports of flying saucer observations. The three colleagues jokingly discussed that the cartoon accurately presented a good hypothesis related to the two stories.

Fig. 7.2 Visitors from other worlds abducting trash cans, possibly to advance their knowledge of humans. Credit: The New Yorker collection 1950, drawn by Alan Dunn, from cartoonbank.com; all rights reserved.

After the three colleagues finished joking, it is reported that Fermi asked: "Where is everybody?". Fermi was referring to extraterrestrial visitors and the fact that many old civilizations could potentially exist with the required technology to visit their whole home galaxy by now. Therefore, we should see evidence of that, but we don't.

The oldest galaxies astronomers have found are believed to have formed only 320 million years after the Big Bang. For example, the galaxy JADES-GS-Z13-0 found using the JWST and reported in April 2023,[5] is believed to have formed when the Universe was only 2% of its present age. If we extrapolate with what we know about life here on Earth, you could argue that such an old galaxy may host civilizations that are up to three times as old as humanity. Using the *Copernican Principle,* which states that humans, Earth, and the solar system are not privileged, and there is nothing special about them, we would expect life to be around in many places in the universe.

Earth's age has been estimated to be 4.5 billion years. Now, imagine planets older than Earth. For the sake of the argument and continuing with our anthropocentric bias, we can narrow such planets to terrestrial planets. For terrestrial planets to be able to form, you need elements such as carbon, silicon, oxygen, iron, and magnesium, among others. These heavier than hydrogen and helium elements are only forged in the interior of stars, then disseminated to the cosmos in supernovae events, resulting from stellar winds.

Therefore, we can infer that terrestrial worlds have not existed from the beginning of the universe. Estimations of the age distribution of terrestrial planets in the universe have determined that the oldest terrestrial planets could be around two billion years older than Earth.[6] If life existed there and intelligence evolved at a similar pace as it did on Earth, then within the last bit of their first 4.5 billion years, life could have become capable of transmitting to the stars and visiting other worlds within their own planetary system. That means that such a life could have had two billion years more than humankind to extend their knowledge and apply that to expand their communications capabilities and explore their own galaxy and the universe.

In less than 100 years, humans achieved sending probes to visit worlds as far as Pluto, located at the edge of the solar system, and currently have rovers exploring Mars and the Moon. Moreover, there are concrete plans to send a spacecraft to collect samples from the plums emanating from the oceans of Enceladus.[7] Imagine how much we will be able to achieve in a billion years if we can avoid self-destruction or successfully defend ourselves against any external threats.

The Fermi Paradox's contradictory nature is centered on the absence of conclusive evidence for the existence of technologically advanced extraterrestrial life, despite the high probability of its existence according to the Copernican Principle. For instance, given that terrestrial planets in the Milky Way could be two billion older than Earth, there is sufficient time for one of those hypothetical civilizations to colonize many worlds.

As organic beings cannot live forever, the use of self-replicating AI machines, or *Von Neumann probes*, as they are known, has been suggested as one of the methods used by an extraterrestrial civilization to extend their presence across and beyond their home galaxy. Proposed by John Von Neumann in 1966,[8] Von Neumann probes are artificial intelligent (AI) machines designed to explore the universe. These probes can self-replicate using resources obtained from planets, satellites, asteroids, and even material in the interstellar medium. The 'parent' probes remain in a given planetary system or region in space, trying to learn as much as possible, while the 'daughter' probes move to the next target in the home galaxy or beyond.

Fig. 7.3 A 'parent' Von Neumann AI probe overseeing a fleet of 'daughter' probes amidst the exploration of a planetary system. Credit: image generated by OpenAI's DALL·E. (can you see the irony?).

Given the current state-of-the-art of AI, it is only expected that a civilization that is older than us by two billion years would have developed such machines. Even if traveling at relatively lower speeds than the speed of light, these probes have had enough time—in order of millions of years—[9] to spread to all stars in our home galaxy, the Milky Way. However, evidence of their existence, such as their propulsion methods and electronic communications to their home planet, is still absent.

Members of the astronomy and science community, and even the public, have come up with ideas of why we don't see such evidence of the existence of extraterrestrial civilizations

These ideas are collectively known as the solutions to the Fermi Paradox. Stephen Webb, a notable physicist and prolific writer, has published a remarkable book that compiles a number of these solutions.

During many interviews, conferences, and talks, Webb has expressed that collecting and documenting these solutions has become a hobby for him. The first edition of his book "If the Universe Is Teeming with Aliens… Where is everybody?" was published in 2002,[10] and contains 50 solutions. The second version, published in 2015,[11] contains 75 solutions.

However, Webb is not the only person who has written about the Fermi Paradox. Countless books, scientific articles, sci-fi stories, and movies have focused on this topic. The discussion is attractive because it explores what could happen to us as a civilization.

Many of the proposed solutions to the Fermi Paradox can be linked to the Drake equation, widely regarded as one of the most famous or second most famous in history, following Einstein's equation describing the mathematical relationship between energy and mass.

The Drake Equation

In his book "A Brief History of Time", Stephen Hawking mentioned: "each equation I included in the book would halve the sales". With that in mind, I have consciously tried not including any equations in this book.

For many, the Drake equation and the Seager Equation, which will be discussed in a moment, are not equations but deemed as probabilistic arguments. Therefore, one could argue that I have succeeded in my purpose of writing an equation-less book (I hope Hawking was right, and that is reflected in the sales of this book). However, for the sake of consistency with popular knowledge, I will keep referring to the Drake and Seager equations as such.

Frank Drake (1930-2022) was an American astrophysicist, astrobiologist, and notable figure in the SETI arena. His "equation", using today's term, was never intended to go "viral". Drake's objective was to stimulate scientific dialogue at the first formal scientific meeting on the topic of SETI in 1961.

The equation summarizes the key concepts scientists consider when estimating, N, or the potential number of technologically advanced alien civilizations in the universe or our own galaxy.

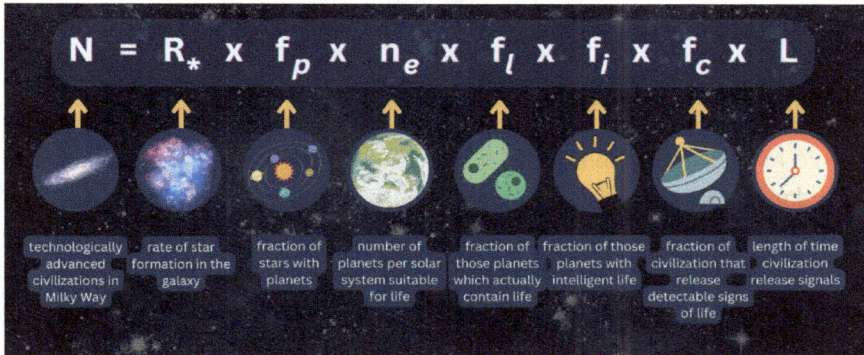

$$N = R_* \times f_p \times n_e \times f_l \times f_i \times f_c \times L$$

| technologically advanced civilizations in Milky Way | rate of star formation in the galaxy | fraction of stars with planets | number of planets per solar system suitable for life | fraction of those planets which actually contain life | fraction of those planets with intelligent life | fraction of civilization that release detectable signs of life | length of time civilization release signals |

Fig. 7.4 The Drake equation. A way to stimulate scientific dialogue about the potential number of advanced technological civilizations in the Universe. Credit: Anne Helmenstine (sciencenotes.org).

Let's explore the terms of the equation and see how they relate to possible solutions to the Fermi Paradox. As usual, focusing on what we know helps narrow our discussion. Hence, narrowing these concepts in terms of the life-as-we-know-it and focusing on our home galaxy, Sun-like stars, and Earth-like planets is a good starting point. While we explore these terms, let's also walk through an exercise where we assign some numbers to them.

Average rate of star formation every year (R^*)

Astronomers have been redefining the methods used to calculate star formation rate. One way to do it is to focus on the local universe. The methods involve the analysis of collected spectra of close-by galaxies. There are two effects that astronomers need to consider: *reddening* and *metallicity*. Reddening in this context refers to the effect that light coming from the stars within a galaxy suffers from its interaction with interstellar dust. Due to the size of the dust particles, blue light waves are absorbed more than red light waves, which makes those stars look redder than what they are. The other aspect to consider is metallicity, which refers to the fraction of heavier elements that are present in a star.

Reddening influences how the luminosities of stars are estimated; therefore, astronomers need to account for such effects when calculating the luminosities of individual stars. By adding up the luminosities of all the stars in a given region, it is possible to estimate the rate of star formation in that region.

On the other hand, as explored in Chapter 1, metallicity provides an alternative to measure when a star was formed. Low metallicity, as in more abundance of hydrogen and helium compared to heavier elements, is an indication that the star must have formed a long time ago. On the contrary, a higher metallicity indicates a younger age for the star. Regions with higher metallicity may have had more star formation in the past, while lower metallicities may indicate more recent star formation. Once levels of metallicity have been established, astronomers use theoretical models and simulations that relate those levels to star formation rates.

Given the amount of research in this area, R^*, or the star formation rate for a region or the universe, is a relatively known number nowadays. This number is given in terms of solar masses. For instance, it has been calculated that, on average, the Milky Way produces between one to two solar masses of stars per year,[12] and this is the number we will use for our exercise. However, the most common stars in the Milky Way are red dwarfs, stars smaller than the Sun. On average, astronomers expect between six to seven new red dwarfs to form in the Milky Way every year.

Going back to the Fermi Paradox, certain regions of the universe may be favored by having a large star rate formation, while others may not be so. A large rate formation increases the probability of stars hosting planets, which can be crucial to the development of life. A solution to why we don't hear anyone out there would be that there are not enough stars in the universe, but we know this is not the case. However, we could say that there are regions of the universe that could host more stars than others, and therefore, those regions could host more potential civilizations.

Fraction of stars that host planets (f_p)

When Frank Drake introduced his equation, astronomers did not know whether planetary systems were ubiquitous across the universe. If that weren't the case, the lack of planets would be a strong solution to the Fermi paradox. After all, if planets were rare, that would explain a lot. However, we now know this is not the case. We extensively discussed in Chapter 3 how astronomers only obtained evidence of other planets outside of the solar system in 1992. Since then, plenty have been detected. By the time the equation was formulated in 1961, astronomers could only make educated guesses to determine the fraction of stars that host planets. However, astronomers now believe that, statistically speaking, every star out there is home to a planetary system. That is, 100% of all the stars in the universe host planets. The Milky Way contains approximately 100 billion stars, and, if we only consider Sun-like stars, the fraction of stars that can host planets in our galaxy will be reduced to 'only' 10%. This still leaves us with approximately 10 billion Sun-like stars. For our exercise, we will assume that the fraction of stars that have planets is 10%.

Average number of planets that can support life (n_e)

A very valid solution for the Fermi Paradox is simply that life does not exist anywhere else. This could happen for a number of reasons, and having planets that are too close (too hot) or too far away (too cold) from their parent star preventing life from flourishing, is definitely a very good reason. This average number of planets that can support life terms, essentially refers to the concept of planets in the *habitable zone* of a star. The habitable zone is the zone at a given distance from the parent star where liquid water can exist on the surface of a planet. Employing the Copernican principle again, in the solar system, two planets reside within the Sun's habitable zone: Earth, and Mars. However, Mars is located at the edge of the Sun's habitable zone and does not enjoy the nice weather we have here on Earth, preventing water from flowing on its liquid form. However, strong evidence suggests that water was indeed flowing at some stage in the past,[13] and could even be present in a few areas of the planet right now.[14] This increases the possibility of finding fossils of any potential life that may have existed on the red planet.[15]

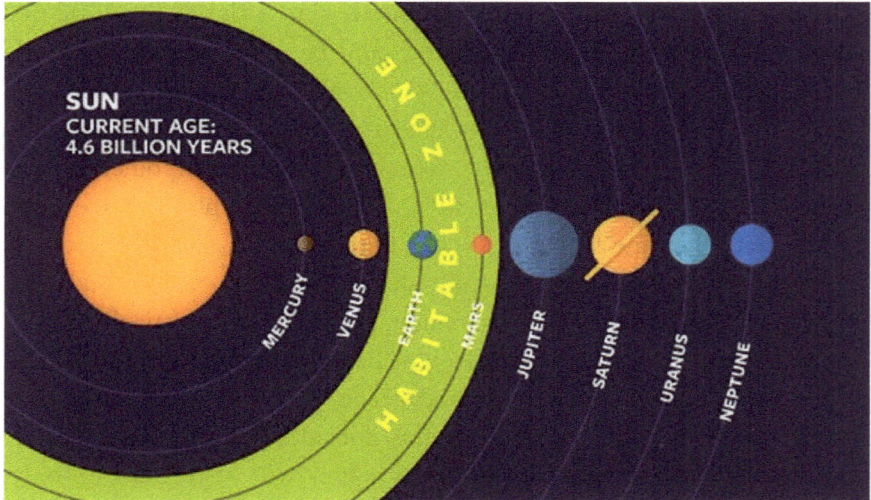

Fig. 7.5 The Sun's habitable zone. Earth and Mars are located in a zone where water could exist on its liquid form in the surface. Credit: Astrobiology.

Therefore, and only considering our definition of habitable zone, we can extrapolate this number and establish that the average number of planets in the solar system that can support life is two.

Fraction of planets on which life can actually appear (f_l)

We are absolutely certain (so far. Ask me again in 10 years, perhaps?), that out of the two planets that are in the Sun's habitable zone, only Earth is the one that currently supports life. Mars could have had the right conditions at some stage (and even Venus), but not anymore. Therefore, in only half, or 50%, of those planets, life emerged and sustained. This, again, gives us some solutions to the Fermi paradox. Life on Earth seems to have flourished due to a very fortunate combination of events: tectonic plates, the size of our Moon, the nature of the solar system formation. But in other places, this might not have been the case. Therefore, other planetary systems can potentially have plenty of planets in the host star's habitable zone, but the conditions may not have been favorable for the development of life.

Now, this parameter of the Drake equation only considers planets and whether they are at the right distance from their host stars. However, the conditions for life to exist are more complicated than just that. Apart from liquid water and building block chemical elements for life, there is also the requirement of having enough energy to allow organisms to perform metabolic processes. In addition, such conditions should be maintained for a considerable time to allow systems and life to evolve. In the solar system, the rocky places where these conditions are more than likely to be met —besides Earth— are not planets but moons.

The reader may remember the 2009 movie Avatar and its 2022 sequel, which portrayed a race of blue tall aliens. The world depicted in these movies was the hypothetical moon Pandora, one of five moons orbiting the gas giant Polyphemus in the Alpha Centauri system.

Fig. 7.6 The Pandora moon in the movie Avatar. The gas giant it orbits can be seen in the background. Credit: Avatar Wiki – Fandom.

Like in the Avatar movies, real moons also orbit gas and ice giants in the solar system. These are worlds where the possibility of the existence of life is real. Let's have a look at these moons.

Jupiter

Europa

Evidence indicates that Europa, a moon of Jupiter and the smallest of the four Galilean moons, has a salty ocean under its icy surface. In 1996, NASA's Galileo spacecraft found that thin plumes of water were vented to space. In addition, Europa's surface is the smoothest of the solar system, suggesting that active processes recycle material from its interior to the surface, creating a potential cycle of organic material. Finally, in September 2023, scientists, using data collected by the JWST, reported carbon dioxide detection on Europa's surface. Further analysis indicates that this carbon likely originated in the subsurface ocean and not as a result of any external source.

The European mission Clipper,[16] which is scheduled to be launched in October 2024, will hopefully shed some light about the moon's habitability and provide initial data that will assist in selecting a landing site for future missions.

Fig. 7.7 A series of dark lines (linae) can be observed on Europa's surface. These lines are potentially the result of a series of eruptions of warm ice. Credit: NASA.

Ganymede

The largest moon of the solar system with a radius of 2,631 kilometers. This is a world that might contain multiple layers of rock, water, and high-pressure ices. This object is the only known moon to have a magnetic field, similar to Earth, that repels dangerous radiation particles, protecting any potential life. For this reason, Ganymede also experiences fluctuating auroras, just like Earth. The fluctuation of these auroras is considered partial evidence of the presence of a large saltwater ocean. But not just any ocean, as it has been estimated that it is 10 times deeper than Earth's oceans.[17]

Fig. 7.8 Image of Ganymede captured by the Juno spacecraft in June 2021. Credit: NASA.

Callisto

The least dense of Jupiter moons. Callisto orbits far away enough to be safe from Jupiter's extreme radiation belts. Scientists are still not sure about the existence of an ocean in this moon, or the depth of such an ocean. The heavily cratered surface is an indication of no geological activity, but the distance from Jupiter, allows Callisto to experience less tidal friction than the other moons. However, due to the low radiation levels, Callisto has long been considered the most suitable place to send possible future crewed missions to establish a permanent base to study the rest of the Jovian system.

Fig. 7.9 The Callisto moon is the most heavily cratered object in the solar system. Credit: NASA.

Saturn

Enceladus

Moving further away from the Sun, we have the tiny Enceladus orbiting Saturn with a radius of only 252.1 kilometers. In this world, plumes of mist emanate from the frozen outer shell and fall back to the surface, making the surface extremely smooth. Enceladus is the most reflective object in the solar system and resembles a big snowball. Geological activity, for which evidence has been piling up recently, along with a potential liquid subsurface ocean, are the key characteristics that lead researchers to believe this is a good place to find life.

Fig. 7.10 Image of Enceladus taken by the Cassini spacecraft at only 25 kilometers of the surface. Credit: NASA.

Titan

The largest moon of Saturn and a place that seems promising for the existence of life. This moon is the only world, other than Earth (that we know of), that has a liquid flowing on its surface. Titan looks remarkably similar to Earth, featuring vast plains, canyonlands, lakes, rain, and clouds. However, the liquid in Titan is not water but methane and ethane, elements that follow a circulation cycle that echoes the water cycles on Earth. Titan's atmosphere is similar to Earth's, with mostly nitrogen but without the oxygen. Despite the likelihood of a subsurface ocean primarily made of water beneath Titan's thick crust of water ice, researchers hypothesize Titan may be home to life that utilizes a chemistry different from the carbon-based life we are used to: methane-based life.[18]

Fig. 7.11 A dark blue dense atmosphere surrounds Titan in this image in false-color captured by the Cassini spacecraft: Credit: NASA/JPL-Caltech/Space Science Institute.

Neptune

Triton

Orbiting the farthest planet from the Sun, Neptune, we find Triton, which revolves around its parent planet in a retrograde motion. Such a characteristic is an indication that this moon is more than likely a captured object from the icy Kuiper Belt. Voyager 2 observed evidence of geysers and lavalike flows on Triton's surface. This strongly suggests the presence of an ocean under a geologically active icy crust and indicates the exchange processes between Triton's surface and subsurface. Telescopic observations of Triton's surface and atmosphere have indicated abundant elemental building blocks, particularly carbon, hydrogen, oxygen, and nitrogen.[19] For all these reasons, Triton is considered one of the most attractive places in the search for life in the solar system.

Fig. 7.12 Global color mosaic of Triton, imaged by Voyager during its flyby of the Neptune system in 1989. Credit: NASA/JPL/USGS.

One fair question that follows is: How is it possible that these moons, so far away from the Sun, can harbor oceans of liquid water? As it happens with planets orbiting their parent stars, the orbits of these moons around their host planets are elliptical, meaning that sometimes they are closer to the planet and sometimes farther away. When the moon is closer to the planet, the enormous gravity of the planet effectively stretches the moon (think of a rugby ball), and once the moon is farther away, the gravitational force is less intense, allowing it to go back to a more circular shape (a basketball). All this stretching and releasing causes heating due to the tidal gravitational forces, *tidal heat*, which allows liquid oceans to exist.[20]

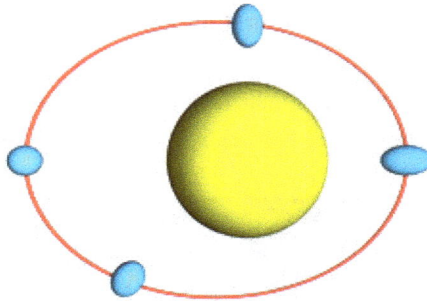

Fig. 7.13 Tidal heating. As the moon orbits closer to its home planet, it gets stretched, only going back to a more circular shape when farther away. Credit: Astrobites. Image by tony Smith.

However, in not all cases, an ocean of liquid water is the result of tidal heating. Some moons, like the case with Callisto, are suspected to have other contributing factors, such as radioactive decay, which also generates heat. An alternative explanation is that the moon could have had ancient impacts that melted ice to form an ocean that has not yet refrozen.[21]

Given the characteristics of these moons in the solar system, one might argue that these worlds should also be considered in this 'fraction of planets on which life can actually appear' parameter of the Drake equation. Moons orbiting exoplanets, or *exomoons* as they are known, could be worlds capable of harboring life. Unfortunately, no strong evidence has been found of the existence of exomoons until now. However, it is likely just a matter of time before astronomers detect the first of them. In fact, Professor David Kipping, a great science communicator and a world authority on exomoons, and his collaborators at the Cool Worlds Lab team at Columbia University, have been trying to confirm the existence of these elusive objects.

In February 2024, Dr Kipping announced that his team has been granted telescope time in the JWST to search for a possible exomoon orbiting the planet Kepler-167 e, a gas giant that orbits the Kepler-167, K-type star. You can learn more about this attempt and more on his YouTube channel,[22] Cool Worlds, which counts more than 800k subscribers up to date.

The fraction of planets that developed intelligent life (f_i)

This parameter refers to the fraction of planets that harbor life that go on to develop intelligent life. The 'Intelligent' term here refers to *human-like* intelligence. On Earth, there are very few, if any, biologists who consider humans the only intelligent race. Dolphins, octopuses, whales, chimpanzees, monkeys, racoons, rats, mice, ravens, and pigeons, just to mention some examples, have exhibited different levels of intelligence. These animals have developed their own languages, can use tools, and seem to have a concept of society.[23] Animals exhibit different levels of reasoning, which is the ability to correlate unrelated experiences into a novel solution for a novel problem. Chimpanzees, for instance, have been observed to solve problems they have not faced before by using available elements. Such as piling boxes and standing on them to reach a suspended banana, even though the chimpanzee had never seen a box before or have seen boxes piled up to reach food. However, animals are still incapable of altering their environment like humans do. They have not developed a space program (not that we know of). Humpback whale songs have estimated to be heard over distances of up to 20 kilometers,[24] but this does not compare to the telecommunications technologies developed by humans.

In summary, plenty of intelligent species can coexist on a planet, but like in our case, only one species has been capable of reaching for the stars.

Therefore, this parameter refers to the fraction of planets that host species exhibiting human-like intelligence, and we have no idea how to calculate this number. If you are optimistic, you can infer that given that humans exist on Earth, every other Earth-like planet out there must have some alien race with human-like intelligence. If you are pessimistic, you can say that human-like intelligence is rare, and we have just been extremely lucky, and you can literally enter whatever number you can think of (including zero). Again, from this parameter, we can infer another solution to the Fermi paradox. Life, not necessarily unicellular life, could be out there, but perhaps, intelligence did not evolve to the point where humans are at the moment. If that is the case, we won't be able to detect them, at least from the point of view of technosignatures.

If you are disappointed about the fact that we have no idea what value to use for this parameter, then you will not be happy with the next parameter either.

However, to prevent you from throwing this book against the wall and to provide a value for this term, we can argue that out of all the planets in the solar system that developed life (the Earth), 100% of them developed human-like intelligence (us). There you go. Based on that assumption, we now have a value for this term.

The fraction of intelligent life that can advertise their existence in space (f_c)

There are no other animals on Earth, again, that we know of, that can manipulate electromagnetic waves to make a video call to communicate with a relative or friend at the other side of the planet.

If you are an extraterrestrial race monitoring Earth and searching for communications in the last 4.5 billion years, you would have probably moved on and selected a different target. That is because humans have only been able to send long-distance messages for less than 200 years.

On May 24, 1844, the first long-distance message was sent using the telegraph system. Thirty years later, the first phone call was made on March 10, 1876. Fast forward less than 150 years to 2023, and you will find that astronauts in the International Space Station will soon be able to have 1Gbps broadband Internet connection.[25] In other words, they will have a better Internet connection than most of us have here on the ground.

So, long range communications by manipulating electromagnetic waves were only developed on Earth in its more recent history. This presents another solution to the Fermi paradox: there could be many civilizations out there, but they either don't have the capability, or the intention to send messages into outer space.

Or perhaps it is physically impossible for some species to develop technology. For example, the concept of an 'oxygen bottleneck' has been proposed as a limit to developing technological capabilities, also known as 'Technosphere'.[26] This hypothesis stems from the fact that combustion on Earth is possible only due to the significant concentration of oxygen in the atmosphere. To put it simply, on a planet with insufficient atmospheric oxygen, combustion could not occur in open-air environments. This would prevent the inhabitants of that planet from using fire to forge tools and, more broadly, from developing metallurgy. For instance, humans could not have produced radio telescopes or other advanced communication methods without such fundamental processes. The same limitation could apply to some extraterrestrial civilizations.

However, there could be civilizations that have indeed developed a Technosphere similar to ours but may have no intention of broadcasting their presence to the universe. This concept is known as The Dark Forest hypothesis.

Imagine you decide to take a walk in the outback, and for whatever reason, you get lost. Nightfall is approaching, and daylight is fading. As you approach a dense, dark forest, you have no idea what lies within. Your instincts tell you to be cautious; you're on the lookout for potential hostile beings or predators. You try to be quiet to avoid revealing your presence to any creatures who might want to harm you or consider you their next meal. You behave this way because it's ingrained in your genes, as a survival strategy. The forest might be silent, too, as its inhabitants could be equally afraid of you —unless they happen to be lions or tigers.

In the context of this parameter of the Drake equation, many civilizations might exist, but they could be afraid of being discovered. Consequently, they refrain from transmitting messages to outer space or take great care to avoid unintentional communication. They might think that if a message is sent to outer space, the recipient could be a hostile race that enjoys preparing meals made from other alien species? While this scenario is quite unlikely, it remains a possibility.

Fig. 7.14 The Dark Forest hypothesis. Civilizations may make the effort of not advertise their existence as they are afraid of encountering a hostile civilization. Credit: Image by Stefan Keller from Pixabay.

On the other hand, for a civilization to be detected, electronic messages will need to be directed to space, either by choice, or as a by-product of their internal communications. The first television signal was broadcast in 1927 by the then W2XCW television station (now known as WRGB) from a facility in New York. A portion of that signal bounced back to space and has been traveling for the last 100 years. This means that any civilization within a 100 light-years sphere radius will be able to detect this signal if they have the adequate equipment. But that is it. If any extraterrestrial civilization were listening before, they would not have detected humans because no one was transmitting. We humans were silent for a long time.

That is another solution to the Fermi paradox. What if many civilizations are out there, but they are still in a pre-communication age? The equivalent to our Stone Age, for instance. What if they never developed the necessary knowledge about electromagnetism, or could not even grasp the basic concepts of physics? They will be silent, not because they want to be silent, but because they don't know how to make noise. So, again, we have no idea about this parameter, and any number is as good as any other one. However, one more time, and similar to what we did for the last parameter, you can also argue that 100% of all the planets in the solar system where intelligent life has developed, intelligent life has been able to advertise their existence to space.

The period that a civilization can communicate across space (L)

How long will a civilization survive? Is a civilization's life span long enough to develop interstellar communication?

In August 1945, the United States detonated two atomic bombs in Japan. The targets were the cities of Hiroshima and Nagasaki. The blasts not only instantly killed more than 140,000 people, mostly civilians, but the subsequent effects of the residual radioactivity took the lives of at least another 80,000 people and left many injured.

Using the same mechanism that is responsible for the release of energy inside of stars, the fusion process (See Chapter 1), the world witnessed how humans had developed the capability to destroy themselves. Such a power has never been seen before in the whole history of humankind.

During the Cuban Missile Crisis on October 27th, 1962, a nuclear-armed submarine from the defunct USSR was around Cuba on international waters. The United States Navy detected the submarine and started dropping signaling charges with the intention they would come to the surface for identification.

The USSR submarine had lost contact with Moscow for several days, and they had no idea whether a war with the U.S. was taking place and whether they needed to respond to what they considered were US attacks.

The USSR submarine was armed with a T-5 nuclear torpedo, which was carrying a destructive 5-kiloton payload. For comparison, the bomb that was detonated in Hiroshima had a 16-kiloton yield, while the one detonated at Nagasaki had a 21-kiloton explosive charge.

Contrary to other USSR submarines that were carrying nuclear weapons, where only the captain was required to authorize a strike, in this submarine, three officers were necessary to authorize an attack. Official documents describe that an argument broke out among the three officers, being the Executive Officer Vasily Aleksandrovich Arkhipov the only one against the launch. Thankfully, Arkhipov won the argument. Having prevented the strike, Akhipov eventually persuaded the commander in charge to surface and await orders from Moscow.[27] Arkhipov's actions effectively prevented a nuclear war, a war that could have ended with millions of casualties at the target regions and cities and massive subsequent deaths because of radiation from the blasts. The consequences of a nuclear war include environmental devastation that could cause massive fires, debris into the atmosphere, and a significant drop in global temperatures due to the reduction in sunlight. Survivors could not probably last longer on the surface and the world as we know it would not exist anymore.

Civilizations out there could be facing, could have faced, or could face similar scenarios. Scientists refer to this type of civilization-destroying events as *filters*. Specifically, they talk about *The Great Filter*, a hypothetical mechanism that prevents life from achieving technologically advanced intelligence.

The Great Filter

There are many examples of what the Great Filter could be. A planet could be sterile due to the inability of life to have a sexual or an asexual reproduction mechanism or the inability to evolve beyond a single-cell structure. A space apocalypse could also have acted as the Great Filter. Events like a nearby supernova explosion, gamma-ray bursts, a close-by black hole, or a large-sized asteroid impact (like the one that may have killed the dinosaurs) are enough to eliminate all traces of life on an unlucky planet. Locally induced processes such as global warming, nuclear winter (because of a nuclear war or accident), or some good intended Artificial Intelligence can also end a civilization in a relatively short period of time. The possibilities are endless, and any of those filters could have wiped out many civilizations out there and could potentially wipe out our own civilization in the future. Have humankind and life on Earth passed all those filters? Are humans now on a trajectory that will allow them to colonize the rest of the solar system and eventually their home galaxy? Or is the great filter still ahead?[28]

The Great Filter

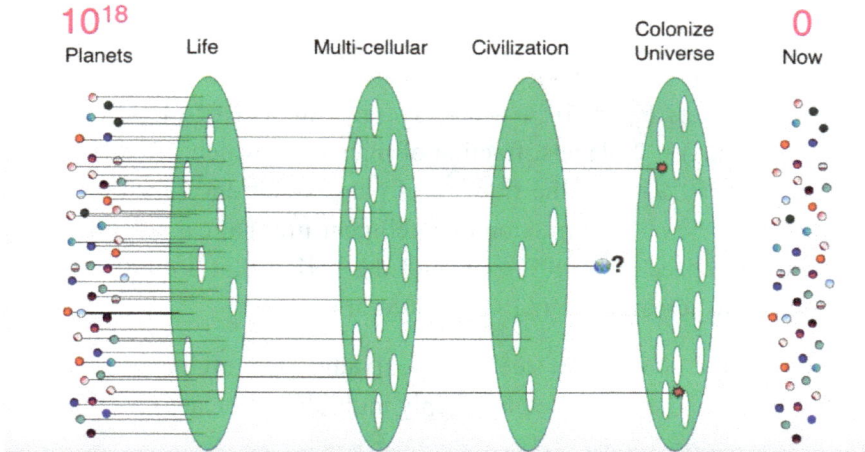

Fig. 7.15 The Great Filter. Significant obstacles that a civilization with human-like intelligence will need to overcome to be detectable. Credit: Illustration by Robin Hanson (2014) The great filter, TEDxLimassol, 4:12.

The Great Filter provides then another set of solutions to the Fermi Paradox. What if, for a given civilization, the filter extinguished them before progressing to the interstellar communication stage? What if the filter did not destroy them but prevented the civilization from achieving the next stage in their technological journey? On Earth, for instance, a large solar storm could completely knock out the power grid and the Internet[29], potentially delaying our technological advances for a while. What if a succession of solar storms disrupts a civilization's technology long enough to halt their path to technological progress? The Great Filter or a significant disruptive one could be upon humans as we can't be certain that such a filter is already behind.

So, what value of L should we use? No idea again. For humanity, pessimistic scientists have suggested 100 years, whereas optimistic scientists suggest one million years. I will take a wild guess and use 10,000 years. Readers are more than welcome to adjust this number according to their beliefs (and mood).

An estimation

We are now able to provide a rough estimate of the number of civilizations in our galaxy. To recapitulate, these are the values we have assigned to all the terms in the Drake equation:

Average rate of star formation every year (R^*) = 2 stars, fraction of stars that host planets (f_p) = 10% = 0.1, average number of planets that can support life (n_e) = 2 planets, fraction of planets on which life can actually appear (f_l) = 50% = 0.5, the fraction of planets that developed intelligent life (f_i) = 100% = 1, the fraction of intelligent life that can advertise their existence in space (f_c)=100% = 1, and the period that a civilization can communicate across space (L) = 10,000 years.

Therefore, the number of technological advanced civilizations in the Milky Way, can be calculated by simply multiplying the above values:

$$N = R^* \times f_p \times n_e \times f_l \times f_i \times f_c \times L$$

$$N = 2 \times 0.1 \times 2 \times 0.5 \times 1 \times 1 \times 10,000$$

$$N = 2,000$$

So, for our example, in our galaxy, 2,000 communicating civilizations should be out there. We have discussed how this number can increase or decrease according to your level of optimism when setting the values of the terms.

In my opinion, the Drake equation serves its purpose. It is just a mechanism to stimulate a discussion about life's existence —or non-existence—

elsewhere. However, not everyone shares the same enthusiasm about this approach. People often state the obvious, indicating the Drake equation is not an equation. I certainly agree with that statement, and I don't believe that was ever Frank Drake's intention. Others call the "equation" speculative and misleading, given the uncertainty of many of its elements. I can definitely relate to those comments. Some of the terms are so unknown that I wonder if we ever get to know their actual values. For this reason, you can play with the equation and enter any number you wish (of course, there is a website for doing just that).[30] Therefore, a solid and robust number for any of the terms is practically impossible.

All these criticisms have prompted researchers and scientists to come up with their own version of the Drake equation. One such version is the Seager equation.

The Seager Equation

The *Biosignature Drake Equation* or *The Seager Equation*[31] is probably the most widely accepted new version of the Drake equation. Proposed by MIT Prof Sara Seager, the equation focuses on biosignature gases instead of technosignatures, and reflects the current and future capabilities to detect such gases. The Seager equation also broadens the potential discovery from only intelligent and communicative life forms, to include any life that interacts with their home planet's atmosphere. The equation also focuses on planets that reside within their home star's habitable zone (HZ).

The equation has six terms and intends to find N, the number of planets with detectable biosignature gases.

Number of stars within the sample (N^*)

Instead of contemplating the entire set of 10 sextillion stars in the universe or all the stars in a galaxy, Prof Seager proposes to only focus on a subset of stars that astronomers can observe. This subset could be, for instance, the number of observable red dwarf stars or the number of observable Sun-like stars. These subsets are specific to a given survey, ground-based telescope, or space observatory. This approach reduces the magnitude of the problem at hand and provides a realistic scenario based on our current capabilities and available data.

The fraction of stars that are quiet (F_Q)

After identifying the set of observable stars as defined in the previous parameter, the next step is to identify the fraction of those with minimal activity. Stars with high solar activity, such as sunspots, present a problem for identifying planets. Sunspots can be confused with a planet transiting the star, leading to a high rate of false positives.

Highly active stars are known to emit large amounts of ultraviolet (UV) light, which can potentially destroy biosignature gases, impairing astronomers' ability to detect them. In addition, high doses of UV radiation can affect life at the cellular level, causing DNA damage. This damage can lead to defective mutations and cells losing their ability to divide and grow. Such conditions are certainly not conducive for life to thrive.

The fraction of stars with rocky planets in the habitable zone (F_{HZ})

This term refers to the number of any potential terrestrial planets orbiting the quiet stars identified in the previous parameter. Again, researchers focus on terrestrial planets, as the only life that they know inhabits a terrestrial world. This parameter also focuses on planets that exist within the habitable zone of a star, as, again, the life we are familiar with requires liquid water in the surface of a planet to exist.

The fraction of rocky planets in the HZ that can be observed (F_O)

There could be many planets out there that are indeed in the HZ of our stars sample. However, will astronomers be able to detect them all? In Chapter 3 we discussed how, for the transit detection method, the orientation of the star-planet system determines its detectability. Contrary to an edge-on planet, or a planet that crosses the observer's line of sight during its transit, if a planet was face-on from the instrument's point of view, astronomers would not be able to detect the dimming that the planet causes when passing in front of its parent star.[32]

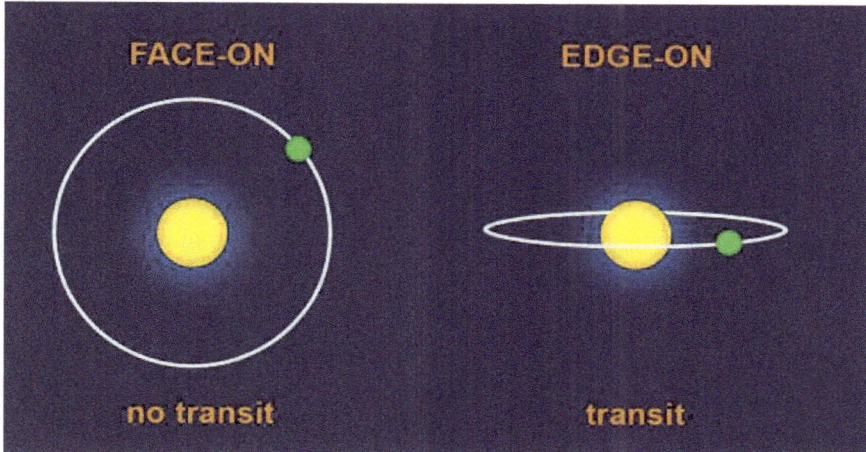

Fig. 7.16 If a planet does not cross the observer's line of sight (face-on), astronomers won't be able to detect its transit, and therefore, they would not know of its existence using the transit method. Credit: Nora Eisner.

As in the case of the transit technique, every exoplanet detection method will also have blind spots where several planets will not be detected due to the planet orbital geometry or technical capabilities.

The fraction of planets that have life (F_L)

Out of all terrestrial planets that orbit within the habitable zone of quiet stars in a given sample of stars, we are now interested in determining the fraction of planets that have life. Similar to the parameter F_l in the Drake equation (Fraction of planets on which life can actually appear), we can talk about the solar system. Earth and Mars are within the habitable zone of their quiet home star, the Sun. Out of those two planets, only one that we know of, harbors life. This means that this parameter for the solar system is 50%. As in the Drake Equation, this parameter is highly specu-lative, and we can only infer its value from our own experience.

The fraction of planets with life that produces a detectable biosignature gas by way of a spectroscopic signature (F_S)

We have discussed how the JWST has enhanced astronomers' ability to characterize the atmospheres of exoplanets, and how this has facilitated the detection of chemical elements present in the atmospheres of certain worlds. For the first time in history, humans have the necessary tools to search for signs of life beyond the solar system. To detect life in these planets, there must be a gas that is unequivocally produced by a biotic process. Moreover, life-produced gases need to accumulate at a level that can be detected in an exoplanet's atmosphere. However, abiotic processes can also generate many gases produced by life on Earth. A prime example is oxygen, considered a favored biosignature gas. However, oxygen can also be produced through the water photodissociation effect. In this chemical reaction, ultraviolet light is absorbed by water molecules, causing hydrogen and oxygen to separate. As hydrogen escapes to space, oxygen accumulates in the atmosphere at detectable levels. Therefore, it is paramount to determine with absolute certainty whether a detected gas is indeed a biosignature. Spoiler alert: scientists will never achieve this 100% certainty. However, models and techniques can be refined to increase the certainty level.

Another estimation

As we can see, the Seager Equation contains fewer unknown and speculative terms than the Drake Equation. As a matter of fact, 71% of the terms are unknown in the Drake Equation, in comparison to only 33% of those terms being speculative in the Seager Equation.

As a result, we can calculate N, or the number of planets with detectable biosignature gases, with a little more confidence. In her seminal work, Prof Seager talks about a couple of estimations. She applies her equation to the Transiting Exoplanet Survey Satellite/James Webb Space Telescope (TESS/JWST) setup.

Professor Seager focuses on the number of M dwarfs stars in the TESS survey, which is around 30,000 ($N^* = 30{,}000$). Out of all these stars, it is estimated that around 60% of those are quiet enough, and therefore, $F_Q = 0.6$.

Using estimations done by other researchers, the fraction of rocky planets in the habitable zone around M stars is about F_{HZ}= 24% or 0.24. The most pessimistic scenario for atmospheres that we can observe in detail for those rocky planets in the HZ of their stars is around 0.1% ($F_O = 0.1\%$ or 0.001).

Therefore, these four terms are well known. The terms that are not well known and require speculation are, of course, the fraction of planets that have life (F_L) and the fraction of planets with life that produce a detectable biosignature gas by way of a spectroscopic signature (F_S). For the first of these terms, F_L, an educated guess will be needed. From our previous discussion in the equivalent Drake's equation term, not all the planets in the habitable zone of a star will support life. We discussed how here, in the solar system, only Earth supports life, in contrast with Mars, a planet also within the Sun's habitable zone. That means that only 50% of the planets within the Sun's habitable zone supports life. We can assign F_L a value of 50% then, based solely on our own experience.

For the last term, F_S, Prof Seager recommends using a value of 50% as well. This speculation is supported by the potentially remotely detectable oxygen present in the atmosphere. Detectable oxygen has existed in Earth's atmosphere for at least half of its life. Multiple lines of evidence indicate that the levels of oxygen in the atmosphere were extremely low before 2.45 billion years ago. However, these oxygen levels became remotely detectable around 2.22 billion years ago.[33]

We now have values for all the six parameters. To summarize, these are the assigned values:

$N^* = 30,000$, $F_Q = 0.6$, $F_{HZ} = 0.24$, $F_O = 0.001$, $F_L = 0.5$, and $F_S = 0.5$. We can now multiply all these numbers to find N, or the number of the number of planets with detectable biosignature gases for the TESS/JWST setup.

A Story of More Than 5000 Worlds

$$N = N^* \times F_Q \times F_{HZ} \times F_O \times F_L \times F_S$$

$$N = 30{,}000 \times 0.6 \times 0.24 \times 0.001 \times 0.5 \times 0.5$$

$$N \sim 1$$

The mathematical symbol '$\sim$' means 'approximately', as the result of our calculation is very close to 1 but not exactly 1.

The obtained result means that by analyzing the collected data for the set of 30,000 target stars using the TESS/JWST setup, the number of planets with detectable signs of life is one.

The importance of the Seager Equation is its ability to serve as a framework to calculate the odds of detecting life for a particular setup or survey. Prof Seager has also calculated the odds for ground-based direct imaging when combining several catalogues, reaching a value for N of approximately three planets.

At the light of these estimations and the pace at which technology is increasing, experts are convinced that the discovery of life outside of Earth is just a matter of time with some of them even indicating that will happen within this decade (2020-2030). The discovery will not necessarily happen at a single point in time as it will require the analysis and classification of large amounts of data and the general consensus of the science community.

On the other hand, some researchers and scientists are not that optimistic about discovering extraterrestrial life based on the detection of a biosignature. With the provocative title '*The futility of Exoplanet Biosignature*',[34] authors Harrison B. Smith and Cole Mathis argue in their research paper that the exercise of unequivocally detecting life by a biosignature is an impossible one. After all, the very same definition of biosignature does not imply the existence of life itself but the identification of a particular process associated with living systems. The authors argue that scientists continue to be on the fence about the definition of life itself and that a full and complete theory of life is still absent.

As we have discussed, biosignatures are ambiguous in nature. There is no easy (or difficult in that regard) way to conclusively claim the discovery of extraterrestrial life due to the detection of a gas related to a biotic process. On the contrary, a technosignature detection, such as the discovery of a SETI signal, will directly lead scientists to conclude the existence of life and, more specifically, human-like intelligent life.

What would happen after we find ET?

What would happen if a SETI signal is indeed detected? Contrary to what Hollywood movies and Internet conspiracy theories tell us, such an event would not be something scientists want to keep a secret. From a scientific point of view, the discoverer or discoverers, would probably be awarded with much recognition, including the Nobel Prize. Beyond preparing for the recognition that such an achievement would bring, the SETI institute has developed the "Declaration of Principles Concerning the Conduct of the Search for Extraterrestrial Intelligence",[35] a compendium of principles that describe the protocol to be followed in case an extraterrestrial intelligence signal is detected. These principles state that once the discovery team has completed the verification process, the team "...shall report this conclusion in a full and complete open manner to the public, the scientific community, and the Secretary General of the United Nations". This means all scientific data will be available to the international scientific community. The actions contemplated in the SETI protocol are common among scientists in reproducing, validating, and confirming scientific findings.

After such a fundamental discovery, either as a result of an accumulative body of evidence or in the form of the detection of a SETI signal, nothing will change in our everyday lives. It is more than likely that such a life (unless detected in our planetary system) will be located very far away from us to the point that we could not reach it within a human lifespan. Most importantly, such a discovery would imply a shift from our current biogeocentric vision to an *astrobiocentric* perspective.

A Story of More Than 5000 Worlds

Astrobiocentrism

Contrary to biocentrism, which views life as unique to Earth, *astrobiocentrism* is a concept that considers the implications for science and humanity if extraterrestrial life is confirmed. As advances and discoveries are being made in the natural sciences with the technology being developed in astrobiology, content and ideas are also being developed on the side of the social sciences and humanities. Let's briefly explore what, in my opinion, are some of the most interesting notions that have been recently introduced:[36, 37]

- **Astrobioethics.** It is a discipline that studies the moral implications and addresses ethical issues related to studying and exploring life in the universe. Bringing humans to Mars and the social responsibility of astrobiologists to society are just some of the examples this discipline deals with.
- **Astrobiosemiotic.** This concept is derived from *semiotics*, which is the study of signs and symbols and their use or interpretation. Astrobiosemiotic focuses on how astrobiologists as interpreters, establish connections between things, between the expression or sign (the biosignature) and the content (the living organism). Astrobiosemiotics' goal is to bring order to the study of biosignatures, enhancing our understanding of life in the universe through the structured interpretation of these signs.
- **Astrotheology.** Most theological reflection is fundamentally geocentric. That is, regardless of theologies often considering the vastness of the universe, they are inherently centered on the emergence of life on Earth and carry a distinctly human creation of meaning. Astrotheology explores fundamental questions such as "Why do we exist?" from an astrobiocentric perspective, where humans are not the center of creation anymore, and anthropocentric and geocentric biases are taken into account to tackle those fundamental questions.

These concepts demonstrate that scholars are getting ready for what is next. Beyond the influence on the work of researchers and scientists, the discovery of extraterrestrial life would profoundly impact our understanding of the universe and how we see ourselves as humanity. New books, Hollywood movies, TV series, and possibly new religions will come up... for sure. Until such definitive proof is detected, we science enthusiasts will keep enjoying the journey and paying attention to the signs (pun intended).

The existence of life beyond Earth remains an enigma, and that's what makes it fun, at least for me. Echoing the profound insights of the remarkable science fiction writer, Arthur C. Clarke:

"Two possibilities exist: either we are alone in the Universe, or we are not. Both are equally terrifying."

Final Remarks

If you made it up to this point, thank you very much for sticking with me. It is customary to finish a book with a final section where the author reaches certain conclusions about what has been discussed. Therefore, I was tempted to use the title "Conclusions" for this final part of the book.

However, I am afraid that for most of the topics we have explored, there are hardly any conclusions we can reach at the moment. Despite all the advancements in science and technology, finding life beyond Earth still seems elusive.

When I was a kid, I was absolutely convinced of the existence of extraterrestrial civilizations and kept looking up to the sky and the TV news, waiting for that moment when we would make first contact. Such convictions were mostly because I could not stop watching all these sci-fi TV shows and movies. In those, aliens would be either our saviors or our executioners. Beings that would show up on our doorstep to deliver some good or bad news (or very bad news, like in the 2005 Hitchhiker's Guide to the Galaxy[1] movie adaptation). However, I have to say that after all these years, I am not that optimistic that this first encounter will happen in our lifetimes… or ever. The reasons for my change of mind are mainly three: i) the immensity of the universe, ii) the conditions for life on Earth to flourish, and iii) the probability that paths collide.

The Universe is Big. Really Big!

The immensity of the universe is something that does not quite fit well in my head yet, and it is more than likely that it never will. My last trip to Colombia from Australia took me nearly three days. The trip, which included many stopovers thanks to the post-COVID economy, included an 11-hour flight from Fiji to Los Angeles, a nearly seven-hour flight from Los Angeles to Panama City, and a close-to-two-hour flight to Santiago de Cali. Such a "wonderful" experience made me think of how big the Earth is. But Earth is so small compared to other structures in the universe. The distances to which we are used on our planet pale in comparison with the vastness of the cosmos. As we discussed throughout the book, these distances are so large that astronomers have come up with units such as the light-year and the parsec.

The most distant human-made object from Earth is NASA's Voyager 1, launched in 1977. The probe has been traveling for nearly 47 years, and its current speed relative to the Sun is 61,200 kilometers per hour. As of February 2024, it is located at 163 times the distance from Earth to the Sun (163 AU). The incredible thing about this is that, technically speaking, Voyager 1 hasn't even left the solar system just yet.

At this speed, Voyager 1 would reach Proxima Centauri, the closest star to the Sun, in approximately 75,000 years. If we want Voyager 1 to adventure outside of the Milky Way, it will take around 441 million years for the probe to reach the edge of the galaxy. So, yeah, it is big. Therefore, the difficulty of achieving interstellar or inter-galactic travel makes me wonder if humans, or any potential extraterrestrial civilization, could cover these large distances. We explored the Fermi paradox and how self-replicating machines (Von Neumann machines) could potentially cover such long distances over thousands or millions of years. If we ever get in contact face-to-face with some extraterrestrial civilization, it will be more than likely with one of such artificial entities.

Is humanity going to survive long enough to welcome one of these explorers? Or will it be our own version of Von Neumann machines that will make that contact?

The Improbability of Complex Life Elsewhere.

On Earth, life is ubiquitous. Simple life—microbial and bacterial—can be found everywhere. Some even deep underground, on the bottoms of oceans, under extreme conditions of cold, heat, and toxicity. Given the affinity of certain microorganisms with extreme environments, these have been dubbed *Extremophiles*, meaning literally "organisms that love the extreme". Extremophiles thrive at temperatures as low as -20 degrees Celsius and as high as 120 degrees Celsius; at pressures of up to 110 MPa, and at extreme acid conditions (pH 0).[2]

This resilience of life in environments that resemble the extreme conditions on many planets of the solar system and exoplanets and the discovery[3] of a large presence of molecules in space that contribute to the construction of amino acids — the basis of genetic material here on Earth — have created an atmosphere of optimism among scientists regarding finding simple life in the universe within the next twenty years.[4] However, complex life — "animal life" or "intelligent life" — is a completely different story.

In the remarkable 2003 book *Rare Earth: Why Complex Life is Uncommon in the Universe*,[5] authors Peter D. Ward and Donald Brownlee laid the case that a confluence of many fortunate circumstances and rare events have made Earth a uniquely habitable environment for complex life. For instance, the large size of our moon contributes to the planet's axial stability, preventing drastic climate changes. The existence of tectonic plates plays a critical role in carbon cycling, climate regulation, and the maintenance of a stable, life-supporting atmosphere over geological timescales.

The solar system's position in the Milky Way galaxy could have also contributed to the success of life on Earth. This concept, known as the *galactic habitable zone (GHZ)*,[6] or *stellar habitable zone*, explores how the position of a planetary system, in our case, 27,000 light-years away from the center of the galaxy, may have played a crucial role in the development of complex life on Earth. The center of the Milky Way exhibits a higher concentration of stars, making it a region where potentially deadly supernovae and gamma-ray bursts may prevent the development of complex life. The centers of galaxies are also known to host supermassive blackholes and active galactic nuclei (See Chapter 1), dangerous sources of high energy that are also detrimental to complex life.

In addition, Earth is at the right location in the galaxy where the abundance of elements has allowed terrestrial planets to form. The metallicity of a host star could play a crucial role in the emergence of complex life on its planets as we explored in Chapter 1. Furthermore, Jupiter has been hypothesized to have served as a shield protecting Earth from comets and asteroids impacts (this argument has not been settled, as it has also been hypothesized that Jupiter could be acting more like a foe rather than a friend).[7] Finally, the Sun is a very stable star with low activity in solar flares, a condition allowing a very stable habitable zone where water can remain liquid. Right atmospheric conditions, a stable planetary system, and the molecular diversity and evolution of lifeforms have allowed a rich diversity of animal life to flourish on Earth.

All this indicates that the existence of complex life in the universe might not be as common as we think. It is true that there are billions of planets out there, but what if the probability of a complex life is so low that not a single planet could host such a life? We don't know; will we ever know?

Finding Each Other

Ten thousand years ago, our ancestors were still living in caves, and their greatest technological achievement was starting a fire to cook their food. What if some alien civilization sent a signal back then from their home planet, and we missed it because we had not developed the technology to receive such a transmission? What if we, as well, are sending signals or observing the home planet of a civilization that is still in an age equivalent to our Stone Age? What if we are the first civilization in the universe? Nearly 14 billion years since the Big Bang, and what are the chances of technological civilizations crossing paths? Being in the right time at the right moment.

What are the chances of two civilizations using equivalent technologies? We explored that some extraterrestrial civilizations could be using neutrinos to communicate. What if, by now, they have completely moved on from radio signals as a way of inter-stellar communications? Many Fermi Paradox solutions also explore the longevity of civilizations. They could have been transmitting, but they are long gone, and our devices won't be aligned again with their transmitters.

All these reasons make me think that either we are indeed alone, or the chances of meeting another human-like intelligence race that is not from Earth is quite small. Should we stop looking, then? Absolutely not. It is true that the chances of finding life out there are quite small, but if we don't look, the chances become zero.

In this golden era we live in, the announcement of the detection of a new planet has somehow become nothing out of the ordinary; but there is nothing trivial about the discovery of a new world. Every new world brings with it new possibilities, fresh ideas, and hope. Every world tells a new story. Stories that collectively improve our understanding of the universe and our place within it.

In the meantime, we, space enthusiasts, will continue to consume science fiction and scientific material and keep looking up at the night sky.

Final Remarks

Like the poster that is hanging in the office of one of the main characters from one of my favorite '90s TV shows reads:

"I Want to Believe"

Acknowledgments

Writing this book allowed me to fulfill a long-held dream and was a truly enjoyable experience. I only have words of appreciation for all the people who helped me along the way.

Many thanks to the numerous readers for their feedback at the different stages of this project. I am extremely thankful for all their time and valuable insights.

Thank you to all my friends who supported me and encouraged me to keep pursuing this dream.

I need to express my gratitude to Paige Lawson, my editor. Her professionalism and on-point edits helped improve the final text notably

I am also extremely grateful to Alexa Eliza, the graphic designer who designed the amazing covers. She was able to translate into images what I could barely express in words.

The awesome formatting for both the digital and print versions is the result of Damian Jackson. Damian captured my requirements amazingly and translated them into the great final layout.

Very appreciative of USQ graduate student Sakhee Bhure. Sakhee was the first person to read the full manuscript. Her science communication skills helped greatly to enrich the text.

Many thanks to Dr Sarah Blunt, who I met virtually during Code/Astro 2022, and Dr Luz Angela Garcia, for their invaluable contributions on the Stars Chapter.

Special thanks to Dr Phil Sutton, not only because his YouTube channel, AstroPhil, helped me clarify concepts on planet migration but also for his remarkable feedback on the Planets Chapter.

Very thankful to my former lecturers at USQ, Dr Brett Addison and Dr Jake Clark, for their extensive contributions in reviewing and fact-checking the chapters on Exoplanets Detection Methods and Exoplanets Classification.

Massive thanks to Dr. Octavio A. Chon Torres, and Northwestern University graduate student Jonathan Roberts, for their wonderful insights, feedback and comments that greatly enhanced the astrobiology and SETI-related chapters.

I need to give a very special acknowledgment to Vicki Anderson, part of the leadership team during my time at Catholic Education network (CEnet). Vicki went above and beyond and gave me sound advice, comments and feedback which contributed greatly to improve the overall readability of the text.

Finally, and most of all, thanks to my lovely wife Claudia, my parents Alejandro and Bertha, and my little brother Julian, for never ceasing to believe in me and for their immense encouragement throughout my life.

Index

Index

Index

Index

Index

Index

Index

Index

Index

Index

Index

Index

About the Author

ALEJANDRO RUIZ RIVERA is a technologist and science enthusiast working for the Queensland Government in Brisbane, Australia, and a member of the ASTROLUCA research group in Cali, Colombia. He obtained his bachelor's degree in Electronics Engineering from Universidad del Valle in Cali, Colombia, in 2003, and his Master's and Doctoral degrees in Telecommunications Engineering from the University of Wollongong in Australia in 2016. Alejandro also graduated with a Master of Science in Astrophysics from the University of Southern Queensland in 2023.

Notes

Preface

1. Sagan, Carl. Cosmos. United States: Random House, 1983.
2. https://www.youtube.com/watch?v=uakLB7Eni2E

1. Stars

1. https://science.nasa.gov/sun/
2. https://imagine.gsfc.nasa.gov/science/objects/milkyway1.html
3. https://theuniversalstory.net/universe-you-here/
4. Hart, M.H., 1978. The evolution of the atmosphere of the Earth. Icarus, 33(1), pp.23-39.
5. Hamacher, D. and Anderson, G.M., 2022. The first astronomers. Allen & Unwin.
6. McKee, C.F. and Ostriker, E.C., 2007. Theory of star formation. *Annu. Rev. Astron. Astrophys.*, *45*, pp.565-687.
7. Weinberg, S. (1977). The First Three Minutes: A Modern View of the Origin of the Universe. New York: Basic Books.
8. Keating, B. (2019). Losing the Nobel Prize: A Story of Cosmology, Ambition, and the Perils of Science's Highest Honor. W. W. Norton & Company.
9. Heng, I. (2021, November 8). Hubble Tension Headache: Clashing Measurements Make the Universe's Expansion a Lingering Mystery. Scientific American. Retrieved from https://www.scientificamerican.com/article/hubble-tension-headache-clashing-measurements-make-the-universes-expansion-a-lingering-mystery/
10. Di Valentino, E., Mena, O., Pan, S., Visinelli, L., Yang, W., Melchiorri, A., Mota, D.F., Riess, A.G. and Silk, J., 2021. In the realm of the Hubble tension—a review of solutions. *Classical and Quantum Gravity*, *38*(15), p.153001.
11. Carroll, B. W., & Ostlie, D. A. (2007). An Introduction to Modern Astrophysics (2nd ed.). San Francisco, CA: Addison-Wesley.
12. https://helenthehare.org.uk/2017/03/19/the-physics-of-ballet/
13. Waters, L.B.F.M. and Waelkens, C., 1998. Herbig ae/be stars. Annual Review of Astronomy and Astrophysics, 36(1), pp.233-266.
14. Bennett, J.O., Donahue, M., Schneider, N. and Voit, M., 2014. *The cosmic perspective* (p. 832). Pearson.
15. http://www.physics.usyd.edu.au/~helenj/LS/LS5-star-birth.pdf
16. Carr, B.J., Bond, J.R. and Arnett, W.D., 1984. Cosmological consequences of Population III stars. Astrophysical Journal, Part 1 (ISSN 0004-637X), vol. 277, Feb. 15, 1984, p. 445-469., 277, pp.445-469.
17. Maiolino, R., Uebler, H., Perna, M., Scholtz, J., D'Eugenio, F., Witten, C., Laporte, N., Witstok, J., Carniani, S., Tacchella, S. and Baker, W., 2023. JWST-JADES. Possible Population III signatures at z= 10.6 in the halo of GN-z11. arXiv preprint arXiv:2306.00953.

Notes

18. Maio, U., Ciardi, B., Dolag, K., Tornatore, L. and Khochfar, S., 2010. The transition from population III to population II-I star formation. Monthly Notices of the Royal Astronomical Society, 407(2), pp.1003-1015.
19. https://www.eso.org/public/australia/news/eso0228/#3
20. Creevey, O.L., Thévenin, F., Boyajian, T.S., Kervella, P., Chiavassa, A., Bigot, L., Mérand, A., Heiter, U., Morel, P., Pichon, B. and Mc Alister, H.A., 2012. Fundamental properties of the Population II fiducial stars HD 122563 and Gmb 1830 from CHARA interferometric observations. Astronomy & Astrophysics, 545, p.A17.
21. Afşar, M., Sneden, C., Frebel, A., Kim, H., Mace, G.N., Kaplan, K.F., Lee, H.I., Oh, H., Oh, J.S., Pak, S. and Park, C., 2016. The Chemical Compositions of Very Metal-poor Stars HD 122563 and HD 140283: A View from the Infrared. The Astrophysical Journal, 819(2), p.103.
22. Fraknoi, A., Morrison, D. and Wolff, S.C., 2018. Astronomy (OpenStax).
23. Shapiro, A.V., Brühl, C., Klingmüller, K., Steil, B., Shapiro, A.I., Witzke, V., Kostogryz, N., Gizon, L., Solanki, S.K. and Lelieveld, J., 2023. Metal-rich stars are less suitable for the evolution of life on their planets. Nature communications, 14(1), p.1893.
24. https://courses.lumenlearning.com/wm-biology1/chapter/reading-spectrums-of-light/
25. Vecteezy.com
26. https://astronomy.com/magazine/glenn-chaple/2021/04/color-coding-stars
27. Tarter, J., 2014, Brown Is Not a Color: 'Introduction of the Term 'Brown Dwarf'', in V Joergens (ed.), *50 Years of Brown Dwarfs*, Astrophysics and Space Science Library, vol 401, Springer International Publishing.
28. Rodriguez, D.R., Zuckerman, B., Melis, C. and Song, I., 2011. The ultra cool brown dwarf companion of WD 0806–661B: age, mass, and formation mechanism. The Astrophysical Journal Letters, 732(2), p.L29.
29. Luhman, K.L., Burgasser, A.J. and Bochanski, J.J., 2011. Discovery of a candidate for the coolest known brown dwarf. The Astrophysical Journal Letters, 730(1), p.L9.
30. https://astrobiology.com/2016/05/hunting-for-hidden-life-on-worlds-orbiting-old-red-stars.html
31. Naoz, S., 2023. Planet swallowed after venturing too close to its star. Nature, 617(7959), pp.38-39.
32. De, K., MacLeod, M., Karambelkar, V., Jencson, J.E., Chakrabarty, D., Conroy, C., Dekany, R., Eilers, A.C., Graham, M.J., Hillenbrand, L.A. and Kara, E., 2023. An infrared transient from a star engulfing a planet. Nature, 617(7959), pp.55-60.
33. Nomoto, K.I., Iwamoto, K. and Kishimoto, N., 1997. Type Ia supernovae: their origin and possible applications in cosmology. Science, 276(5317), pp.1378-1382.
34. Perlmutter, S., 2000, "Supernovae, dark energy, and the accelerating universe: The status of the cosmological parameters", International Journal of Modern Physics A, vol. 15, pp. 715-739.
35. Riess, A.G., et. al., 1998, "Observational evidence from supernovae for an accelerating universe and a cosmological constant", The Astronomical Journal, vol. 116, no. 3, pp.1009-1045.
36. Thielemann, F.K., Eichler, M., Panov, I.V. and Wehmeyer, B., 2017. Neutron star mergers and nucleosynthesis of heavy elements. Annual Review of Nuclear and Particle Science, 67, pp.253-274.
37. https://svs.gsfc.nasa.gov/13873
38. https://science.nasa.gov/universe/neutron-stars-are-weird/

39. Proudfoot, B., 2021. She changed astronomy forever. He won the Nobel Prize for it. *The New York Times, 27*.
40. S. Sheikh, D. Pines, P. Ray, K. Wood, M. Lovellette, and M. Wolff, "Spacecraft navigation using x-ray pulsars," Journal of Guidance Control and Dynamics, vol. 29, no. 1, pp. 49–63, Jan 2006
41. Y. Wang, W. Zheng, S. Sun, and L. Li, "X-ray pulsar-based navigation using time-differenced measurement," Aerospace Science and Technology, vol. 36, pp. 27–35, Jul 2014.
42. Zhongming, Z., Linong, L., Xiaona, Y., Wangqiang, Z. and Wei, L., 2019. In-orbit Demonstration of X-Ray Pulsar Navigation with the Insight-HXMT Satellite.
43. The Neutron Star Interior Composition Explorer Mission: https://heasarc.gsfc.nasa.gov/docs/nicer/
44. Kaur, T., Blair, D., Moschilla, J., Stannard, W. and Zadnik, M., 2017. Teaching Einsteinian physics at schools: part 1, models and analogies for relativity. *Physics Education, 52*(6), p.065012.
45. Webster, B.L., Murdin, P. and Bolton, C.T., 1979. The Discovery of a Candidate Black Hole. In *A Source Book in Astronomy and Astrophysics, 1900–1975* (pp. 460-465). Harvard University Press.
46. El-Badry, K., Rix, H.W., Quataert, E., Howard, A.W., Isaacson, H., Fuller, J., Hawkins, K., Breivik, K., Wong, K.W., Rodriguez, A.C. and Conroy, C., 2023. A Sun-like star orbiting a black hole. *Monthly Notices of the Royal Astronomical Society, 518*(1), pp.1057-1085.
47. Patel, R., Roachell, B., Caino-Lores, S., Ketron, R., Leonard, J., Tan, N., Brown, D., Deelman, E. and Taufer, M., 2022. Reproducibility of the first image of a black hole in the galaxy M87 from the event horizon telescope (EHT) collaboration. arXiv preprint arXiv:2205.10267.
48. Pounds, K.A., Nixon, C.J., Lobban, A. and King, A.R., 2018. An ultrafast inflow in the luminous Seyfert PG1211+ 143. Monthly Notices of the Royal Astronomical Society, 481(2), pp.1832-1838.
49. Nightingale, J.W., Smith, R.J., He, Q., O'Riordan, C.M., Kegerreis, J.A., Amvrosiadis, A., Edge, A.C., Etherington, A., Hayes, R.G., Kelly, A. and Lucey, J.R., 2023. Abell 1201: detection of an ultramassive black hole in a strong gravitational lens. Monthly Notices of the Royal Astronomical Society, 521(3), pp.3298-3322.
50. Volonteri, M., Habouzit, M. and Colpi, M., 2021. The origins of massive black holes. *Nature Reviews Physics, 3*(11), pp.732-743.

2. Planets

1. https://www.iau.org/public/themes/pluto/
2. Christophe, B., Spilker, L.J., Anderson, J.D., André, N., Asmar, S.W., Aurnou, J., Banfield, D., Barucci, A., Bertolami, O., Bingham, R. and Brown, P., 2012. OSS (Outer Solar System): a fundamental and planetary physics mission to Neptune, Triton and the Kuiper Belt. Experimental Astronomy, 34, pp.203-242.
3. New Horizons mission: https://science.nasa.gov/mission/new-horizons/
4. Stern, S.A., Bagenal, F., Ennico, K., Gladstone, G.R., Grundy, W.M., McKinnon, W.B., Moore, J.M., Olkin, C.B., Spencer, J.R., Weaver, H.A. and Young, L.A., 2015. The Pluto system: Initial results from its exploration by New Horizons. Science, 350(6258), p.aad1815.

Notes

5. Batygin, K. and Brown, M.E., 2016. Evidence for a distant giant planet in the solar system. The Astronomical Journal, 151(2), p.22.
6. Batygin, K., Morbidelli, A., Brown, M.E. and Nesvorny, D., 2024. Generation of Low-Inclination, Neptune-Crossing TNOs by Planet Nine. arXiv preprint arXiv:2404.11594.
7. Andrews, S.M., Huang, J., Pérez, L.M., Isella, A., Dullemond, C.P., Kurtovic, N.T., Guzmán, V.V., Carpenter, J.M., Wilner, D.J., Zhang, S. and Zhu, Z., 2018. The disk substructures at high angular resolution project (DSHARP). I. Motivation, sample, calibration, and overview. The Astrophysical Journal Letters, 869(2), p.L41.
8. Disk Substructures at High Angular Resolution Project (DSHARP): https://alma science.eso.org/almadata/lp/DSHARP/
9. How old is it -The Solar System: https://www.youtube.com/watch?v=8N7VzHScNsE
10. Spencer, J.R., Stern, S.A., Moore, J.M., Weaver, H.A., Singer, K.N., Olkin, C.B., Verbiscer, A.J., McKinnon, W.B., Parker, J.W., Beyer, R.A. and Keane, J.T., 2020. The geology and geophysics of Kuiper Belt object (486958) Arrokoth. Science, 367(6481), p.eaay3999.
11. Klahr, H. and Bodenheimer, P., 2006. Formation of giant planets by concurrent accretion of solids and gas inside an anticyclonic vortex. The Astrophysical Journal, 639(1), p.432.
12. https://www.youtube.com/watch?v=BGPBNeTFXZk&t=2s
13. https://youtu.be/6AcakIR7MRk?si=HKxJ1CsidQrYH2T-
14. Plavchan, P. and Bilinski, C., 2013. Stars do not Eat Their Young Migrating Planets: Empirical Constraints on Planet Migration Halting Mechanisms. The Astrophysical Journal, 769(2), p.86.
15. Gáspár, A., Wolff, S.G., Rieke, G.H. et al. Spatially resolved imaging of the inner Fomalhaut disk using JWST/MIRI. Nat Astron (2023). https://doi.org/10.1038/s41550-023-01962-6
16. https://astronomy.swin.edu.au/cosmos/c/Cometary+Gas+Tail
17. https://www.wired.com/2013/03/why-does-a-comet-have-a-tail/
18. https://solarsystem.nasa.gov/stardust/comets/hb.html
19. Strøm, P.A., Bodewits, D., Knight, M.M., Kiefer, F., Jones, G.H., Kral, Q., Matrà, L., Bodman, E., Capria, M.T., Cleeves, I. and Fitzsimmons, A., 2020. Exocomets from a solar system perspective. Publications of the Astronomical Society of the Pacific, 132(1016), p.101001.
20. Janson, M., Patel, J., Ringqvist, S.C., Lu, C., Rebollido, I., Lichtenberg, T., Brandeker, A., Angerhausen, D. and Noack, L., 2023. Imaging of exocomets with infrared interferometry. Astronomy & Astrophysics, 671, p.A114.
21. Hartmann, W.K., and Davis, D.L., Satellite-sized Planetesimals and Lunar Origin, Icarus 24 (1975), pp. 504-515
22. Mackenzie, D., 2003, 'The big splat or how our Moon came to be', John Wiley & Sons, Inc, Hoboken, New Jersey.
23. The Hoba Meteorite: https://www.info-namibia.com/activities-and-places-of-interest/otavi/hoba-meteorite
24. Chiarenza, A.A., Farnsworth, A., Mannion, P.D., Lunt, D.J., Valdes, P.J., Morgan, J.V. and Allison, P.A., 2020. Asteroid impact, not volcanism, caused the end-Cretaceous dinosaur extinction. Proceedings of the National
25. Schulte, P., Alegret, L., Arenillas, I., Arz, J.A., Barton, P.J., Bown, P.R., Bralower, T.J., Christeson, G.L., Claeys, P., Cockell, C.S. and Collins, G.S., 2010. The Chicxulub asteroid impact and mass extinction at the Cretaceous-Paleogene boundary. Science, 327(5970), pp.1214-1218.

Notes

26. Longo, G., 2007. The Tunguska event. In Comet/Asteroid Impacts and Human Society: An Interdisciplinary Approach (pp. 303-330). Berlin, Heidelberg: Springer Berlin Heidelberg.
27. https://www.un.org/en/observances/asteroid-day
28. https://science.nasa.gov/mission/dart/

3. Exoplanets Detection Methods – First Part

1. Bruno, G., The Heroic Enthusiasts, translated by Williams, L., London, 1887 (1548).
2. Urone, P.P., and Hinrichs, R., 1998, 'College Physics', openstax, Houston, Texas.
3. Struve, O., 1952. Proposal for a project of high-precision stellar radial velocity work. The Observatory,72, pp.199-200.
4. Wolszczan, A., 1994. Confirmation of Earth-mass planets orbiting the millisecond pulsar PSR B1257+ 12. Science, 264(5158), pp.538-542.
5. https://exoplanetarchive.ipac.caltech.edu/docs/counts_detail.html
6. https://spaceplace.nasa.gov/barycenter/en/
7. https://www.eso.org/public/teles-instr/lasilla/36/nirps/
8. https://www.cfa.harvard.edu/facilities-technology/telescopes-instruments/high-accu racy-radial-velocity-planet-searcher
9. http://exoplanets.astro.yale.edu/instrumentation/expres.php
10. https://online-learning-college.com/knowledge-hub/gcses/gcse-physics-help/doppler-effect/
11. https://www.anisotropela.dk/encyclo/redshift.html
12. https://exoplanetarchive.ipac.caltech.edu/docs/counts_detail.html
13. Zeilik, M. and Gregory, S., 1998. Introductory astronomy and astrophysics.
14. Baranne, A., Queloz, D., Mayor, M., Adrianzyk, G., Knispel, G., Kohler, D., Lacroix, D., Meunier, J.P., Rimbaud, G. and Vin, A., 1996, 'ELODIE: A spectrograph for accurate radial velocity measurements', Astronomy and Astrophysics Supplement Series, vol. 119, no. 2, pp. 373-390.
15. Wei, J., 2018, 'A Survey of Exoplanetary Detection Techniques', arXiv e-prints, pp. arXiv-1805.
16. Fischer, D.A., Howard, A.W., Laughlin, G.P., Macintosh, B., Mahadevan, S., Sahlmann, J. and Yee, J.C., 2015, 'Exoplanet detection techniques', arXiv preprint arXiv:1505.06869.
17. Fischer, D.A., Howard, A.W., Laughlin, G.P., Macintosh, B., Mahadevan, S., Sahlmann, J. and Yee, J.C., 2015, 'Exoplanet detection techniques', arXiv preprint arXiv:1505.06869.
18. Fabian, S.T., Sondhi, Y., Allen, P.E., Theobald, J.C. and Lin, H.T., 2024. Why flying insects gather at artificial light. Nature Communications, 15(1), p.689.
19. https://spacemath.gsfc.nasa.gov/earth/4Page28.pdf
20. Jara-Maldonado, M., Alarcon-Aquino, V., Rosas-Romero, R., Starostenko, O. and Ramirez-Cortes, J.M., 2020. 'Transiting exoplanet discovery using machine learning techniques: a survey'. Earth Science Informatics, vol. 13, no. 3, pp. 573-600.
21. Pollacco, D.L., Skillen, I., Cameron, A.C., Christian, D.J., Hellier, C., Irwin, J., Lister, T.A., Street, R.A., West, R.G., Anderson, D. ,Clarkson, W.I., Deeg, H., Enoch, B., Evans, A., Fitzsimmons, A., Haswell, C.A., Hodgkin, S., Horne, K., Kane, S.R., Keenan, F.P., Maxted, P.F.L., Norton, A.J., Osborne, J., Parley, N.R., Ryans, R.S.I., Smalley, B., Wheatley, P.J. and Wilson, D.M., 2006, 'The WASP project and the

SuperWASP cameras', Publications of the Astronomical Society of the Pacific, vol. 118, no. 848, pp. 1407-1418

22. https://cfa.harvard.edu/facilities-technology/telescopes-instruments/hungarian-made-automated-telescope-network

23. Jenkins, J.M., Caldwell, D.A., Chandrasekaran, H., Twicken, J.D., Bryson, S.T., Quintana, E.V., Clarke, B.D., Li, J., Allen, C., Tenenbaum, P. and Wu, H., 2010, 'Overview of the Kepler science processing pipeline', The Astrophysical Journal Letters, vol. 713, no. 2, pp. L87-L91.

24. Howard, A.W., 2015, 'Transiting Exoplanet Survey Satellite', Journal of Astronomical Telescopes, Instruments, and Systems, vol. 1, no. 1, pp. 0140031-01400310.

25. Howell, S.B., Sobeck, C., Haas, M., Still, M., Barclay, T., Mullally, F., Troeltzsch, J., Aigrain, S., Bryson, S.T., Caldwell, D. and Chaplin, W.J., 2014, 'The K2 mission: characterization and early results', Publications of the astronomical Society of the Pacific, vol. 126, no. 938, pp. 398-408.

26. Perryman, M., 2018, 'the exoplanet handbook', 2nd edn, Cambridge University Press, Cambridge.

27. Reipurth, B., Jewitt, D. and Keil, K. eds., 2007. Protostars and planets V. University of Arizona Press.

28. https://exoplanetarchive.ipac.caltech.edu/index.html

4. Exoplanets Detection Methods – Second Part

1. https://exoplanetarchive.ipac.caltech.edu/docs/counts_detail.html

2. Seidelmann, P. Kenneth., Kovalevsky, Jean. Fundamentals of Astrometry. N.p.:Cambridge University Press, 2004.

3. https://www.cupix.com/news-info/how-to-take-360deg-photos-at-a-fixed-npp-and-manually-stitch-them

4. Fischer, D.A., Howard, A.W., Laughlin, G.P., Macintosh, B., Mahadevan, S., Sahlmann, J. and Yee, J.C., 2015. Exoplanet detection techniques. arXiv preprint arXiv:1505.06869.

5. Sahlmann, J., Lazorenko, P.F., Ségransan, D., Martín, E.L., Queloz, D., Mayor, M. and Udry, S., 2013. Astrometric orbit of a low-mass companion to an ultracool dwarf. Astronomy & Astrophysics, 556, p.A133.

6. Curiel, S., Ortiz-León, G.N., Mioduszewski, A.J. and Sanchez-Bermudez, J., 2022. 3D orbital architecture of a dwarf binary system and its planetary companion. The Astronomical Journal, 164(3), p.93.

7. Currie, T., Brandt, G.M., Brandt, T.D., Lacy, B., Burrows, A., Guyon, O., Tamura, M., Liu, R.Y., Sagynbayeva, S., Tobin, T. and Chilcote, J., 2022. Direct Imaging and Astrometric Discovery of a Superjovian Planet Orbiting an Accelerating Star. arXiv preprint arXiv:2212.00034.

8. Perryman, M., 2018. The exoplanet handbook. Cambridge university press.

9. https://www.youtube.com/watch?v=6viGNvZiscc&list=PLzgLe2CcYDfJILbYLNJBopeX1V0UXLmTG&index=1&t=24s

10. Chauvin, G., Lagrange, A.M., Dumas, C., Zuckerman, B., Mouillet, D., Song, I., Beuzit, J.L. and Lowrance, P., 2004. A giant planet candidate near a young Brown Dwarf-direct vlt/naco observations using ir wavefront sensing. Astronomy & Astrophysics, 425(2), pp.L29-L32.

11. https://www.eso.org/public/spain/news/eso0428/

Notes

12. https://www.nasa.gov/image-feature/goddard/2022/hubble-spies-sparkling-spray-of-stars-in-ngc-2660

13. Carter, A.L., Hinkley, S., Kammerer, J., Skemer, A., Biller, B.A., Leisenring, J.M., Millar-Blanchaer, M.A., Petrus, S., Stone, J.M., Ward-Duong, K. and Wang, J.J., 2022. The JWST Early Release Science Program for Direct Observations of Exoplanetary Systems I: High Contrast Imaging of the Exoplanet HIP 65426 b from 2-16_mu m. arXiv preprint arXiv:2208.14990.

14. https://webbtelescope.org/contents/media/images/01GBT1E93YV7YND5MF S1603FWJ

15. https://www.jpl.nasa.gov/missions/the-nancy-grace-roman-space-telescope

16. Spitzer, L., 1962. The beginnings and future of space astronomy. American Scientist, 50(3), pp.473-484.

17. Cash, W., Hyde, T., Polidan, R. and Glassman, T., Starshade Technology Development.

18. Seager, S., Turnbull, M., Sparks, W., Thomson, M., Shaklan, S.B., Roberge, A., Kuchner, M., Kasdin, N.J., Domagal-Goldman, S., Cash, W. and Warfield, K., 2015, September. The Exo-S probe class starshade mission. In Techniques and Instrumentation for Detection of Exoplanets VII (Vol. 9605, pp. 273-290). SPIE.

19. Janson, M., Henning, T., Quanz, S.P., Asensio-Torres, R., Buchhave, L., Krause, O., Palle, E. and Brandeker, A., 2021. Occulter to earth: prospects for studying earth-like planets with the E-ELT and a space-based occulter. Experimental Astronomy, pp.1-14.

20. Hewitt, J.N., Turner, E.L., Schneider, D.P., Burke, B.F., Langston, G.I. and Lawrence, C.R., 1988. Unusual radio source MG1131+ 0456: a possible Einstein ring. *Nature*, *333*(6173), pp.537-540.

21. Bennett, D.P., Anderson, J., Bond, I.A., Udalski, A. and Gould, A., 2006. Identification of the OGLE-2003-BLG-235/MOA-2003-BLG-53 planetary host star. *The Astrophysical Journal*, *647*(2), p.L171.

22. http://ogle.astrouw.edu.pl/

23. https://www.daviddarling.info/encyclopedia/P/PLANET.html

24. https://kmtnet.kasi.re.kr/kmtnet-eng/

25. https://www.universetoday.com/138141/gravitational-microlensing-method/

26. Turyshev, S.G., 2017. Wave-theoretical description of the solar gravitational lens. *Physical Review D*, *95*(8), p.084041.

27. arXiv:2303.14917 [astro-ph.EP]

28. Tomiki, A., Mimasu, Y., Ogawa, N., Matsumoto, J. and ITO, T., Orbit Determination for Long-term Prediction of Solar Power Sail Demonstrator IKAROS.

29. https://en.wikipedia.org/wiki/IKAROS#/media/File:IKAROS_IAC_2010.jpg

30. Silvotti, R., Schuh, S., Janulis, R., Solheim, J.E., Bernabei, S., Østensen, R., Oswalt, T.D., Bruni, I., Gualandi, R., Bonanno, A. and Vauclair, G., 2007. A giant planet orbiting the 'extreme horizontal branch'star V 391 Pegasi. Nature, 449(7159), pp.189-191.

31. Murphy, S.J., Bedding, T.R. and Shibahashi, H., 2016. A planet in an 840 day orbit around a Kepler main-sequence A star found from phase modulation of its pulsations. The Astrophysical Journal Letters, 827(1), p.L17.

32. https://www.youtube.com/watch?v=rqQ1xKsNIQE

33. Heller, R., Hippke, M., Placek, B., Angerhausen, D. and Agol, E., 2016. Predictable patterns in planetary transit timing variations and transit duration variations due to exomoons. Astronomy & Astrophysics, 591, p.A67.

34. Adibekyan, V., De Laverny, P., Recio-Blanco, A., Sousa, S.G., Delgado-Mena, E., Kordopatis, G., Ferreira, A.C.S., Santos, N.C., Hakobyan, A.A. and Tsantaki, M.,

Notes

2018. The AMBRE project: searching for the closest solar siblings. Astronomy & Astrophysics, 619, p.A130.

35. Potter, S.B., Romero-Colmenero, E., Ramsay, G., Crawford, S., Gulbis, A., Barway, S., Zietsman, E., Kotze, M., Buckley, D.A., O'Donoghue, D. and Siegmund, O.H.W., 2011. Possible detection of two giant extrasolar planets orbiting the eclipsing polar UZ Fornacis. Monthly Notices of the Royal Astronomical Society, 416(3), pp.2202-2211.

36. Applegate, J.H., 1992. A mechanism for orbital period modulation in close binaries. Astrophysical Journal, Part 1 (ISSN 0004-637X), vol. 385, Feb. 1, 1992, p. 621-629., 385, pp.621-629.

37. https://www.britannica.com/science/lunar-phases

38. https://www.instagram.com/seanorphoto/?hl=en

39. Kimball, Donald Stevens. "A study of the aurora of 1859." (1960).

40. Vida, K., Seli, B., Szklenár, T., Kriskovics, L., Görgei, A. and Kővári, Z., 2024. Detecting coronal mass ejections with machine learning methods. arXiv preprint arXiv:2401.07588.

41. Faigler, S., Tal-Or, L., Mazeh, T., Latham, D.W. and Buchhave, L.A., 2013. Beer analysis of kepler and corot light curves. i. discovery of kepler-76b: A hot jupiter with evidence for superrotation. The Astrophysical Journal, 771(1), p.26.

42. Pinte, C., van Der Plas, G., Ménard, F., Price, D.J., Christiaens, V., Hill, T., Mentiplay, D., Ginski, C., Choquet, E., Boehler, Y. and Duchêne, G., 2019. Kinematic detection of a planet carving a gap in a protoplanetary disk. *Nature Astronomy*, *3*(12), pp.1109-1114.

5. Exoplanets Classification

1. Latham, D.W., Mazeh, T., Stefanik, R.P., Mayor, M. and Burki, G., 1989. The unseen companion of HD114762: a probable Brown Dwarf. Nature, 339(6219), pp.38-40.

2. Kiefer, F., 2019. Determining the mass of the planetary candidate HD 114762 b using Gaia. Astronomy & Astrophysics, 632, p.L9.

3. Gaudi, B.S., Stassun, K.G., Collins, K.A., Beatty, T.G., Zhou, G., Latham, D.W., Bieryla, A., Eastman, J.D., Siverd, R.J., Crepp, J.R. and Gonzales, E.J., 2017. A giant planet undergoing extreme-ultraviolet irradiation by its hot massive-star host. Nature, 546(7659), pp.514-518.

4. Mankovich, C.R. and Fuller, J., 2021. A diffuse core in Saturn revealed by ring seismology. Nature Astronomy, 5(11), pp.1103-1109.

5. Meyer, M.R., Hillenbrand, L.A., Backman, D., Beckwith, S., Bouwman, J., Brooke, T., Carpenter, J., Cohen, M., Cortes, S., Crockett, N. and Gorti, U., 2006. The formation and evolution of planetary systems: Placing our solar system in context with Spitzer. Publications of the Astronomical Society of the Pacific, 118(850), p.1690.

6. Sánchez-Lavega, A., García-Melendo, E., Legarreta, J., Miró, A., Soria, M. and Ahrens-Velásquez, K., 2024. The Origin of Jupiter's Great Red Spot. Geophysical Research Letters, 51(12), p.e2024GL108993.

7. Irwin, P.G., Teanby, N.A., Fletcher, L.N., Toledo, D., Orton, G.S., Wong, M.H., Roman, M.T., Pérez-Hoyos, S., James, A. and Dobinson, J., 2022. Hazy blue worlds: A holistic aerosol model for Uranus and Neptune, including dark spots. Journal of Geophysical Research: Planets, 127(6), p.e2022JE007189.

8. Frelikh, R. and Murray-Clay, R.A., 2017. The formation of Uranus and Neptune: fine-tuning in core accretion. The Astronomical Journal, 154(3), p.98.

Notes

9. Haywood, R.D., Vanderburg, A., Mortier, A., Giles, H.A., Lopez-Morales, M., Lopez, E.D., Malavolta, L., Charbonneau, D., Cameron, A.C., Coughlin, J.L. and Dressing, C.D., 2018. An Accurate Mass Determination for Kepler-1655b, a Moderately Irradiated World with a Significant Volatile Envelope. The Astronomical Journal, 155(5), p.203.

10. Cassan, A., Kubas, D., Beaulieu, J.P., Dominik, M., Horne, K., Greenhill, J., Wambsganss, J., Menzies, J., Williams, A., Jørgensen, U.G. and Udalski, A., 2012. One or more bound planets per Milky Way star from microlensing observations. Nature, 481(7380), pp.167-169.

11. https://www.jpl.nasa.gov/news/study-shows-our-galaxy-has-at-least-100-billion-planets

12. https://exoplanets.nasa.gov/news/1419/nasa-telescope-reveals-largest-batch-of-earth-size-habitable-zone-planets-around-single-star/

13. Zieba, S., Kreidberg, L., Ducrot, E. et al. No thick carbon dioxide atmosphere on the rocky exoplanet TRAPPIST-1 c. Nature (2023). https://doi.org/10.1038/s41586-023-06232-z

14. Greene, T.P., Bell, T.J., Ducrot, E., Dyrek, A., Lagage, P.O. and Fortney, J.J., 2023. Thermal emission from the Earth-sized exoplanet TRAPPIST-1 b using JWST. Nature, 618(7963), pp.39-42.

15. Rivera, E.J., Lissauer, J.J., Butler, R.P., Marcy, G.W., Vogt, S.S., Fischer, D.A., Brown, T.M., Laughlin, G. and Henry, G.W., 2005. A~ 7.5 M⊕ planet orbiting the nearby star, GJ 876. The Astrophysical Journal, 634(1), p.625.

16. Jenkins, J.M., Twicken, J.D., Batalha, N.M., Caldwell, D.A., Cochran, W.D., Endl, M., Latham, D.W., Esquerdo, G.A., Seader, S., Bieryla, A. and Petigura, E., 2015. Discovery and validation of Kepler-452b: a 1.6 R⊕ super Earth exoplanet in the habitable zone of a G2 star. The Astronomical Journal, 150(2), p.56.

17. Mishra, L., Alibert, Y., Udry, S. and Mordasini, C., 2023. Framework for the architecture of exoplanetary systems-I. Four classes of planetary system architecture. Astronomy & Astrophysics, 670, p.A68.

18. https://www.youtube.com/watch?v=bv_hWreRJDU

19. https://exodashboard.streamlit.app/

20. https://streamlit.io/

21. Hebb, L., Collier-Cameron, A., Triaud, A.H.M.J., Lister, T.A., Smalley, B., Maxted, P.F.L., Hellier, C., Anderson, D.R., Pollacco, D., Gillon, M. and Queloz, D., 2009. WASP-19b: the shortest period transiting exoplanet yet discovered. The Astrophysical Journal, 708(1), p.224.

22. Charnoz, S., Dones, L., Esposito, L.W., Estrada, P.R. and Hedman, M.M., 2009. Origin and evolution of Saturn's ring system. Saturn from Cassini-Huygens, pp.537-575.

23. Teodoro, Luís FA, Jacob A. Kegerreis, Paul R. Estrada, Matija Ćuk, Vincent R. Eke, Jeffrey N. Cuzzi, Richard J. Massey, and Thomas D. Sandnes. "A recent impact origin of Saturn's rings and mid-sized moons."The Astrophysical Journal 955, no. 2 (2023): 137.

24. Black, B.A. and Mittal, T., 2015. The demise of Phobos and development of a Martian ring system. Nature Geoscience, 8(12), pp.913-917.

25. https://www.flickr.com/photos/192271236@N03/53635851891/

26. Anglada-Escudé, G., Amado, P.J., Barnes, J., Berdiñas, Z.M., Butler, R.P., Coleman, G.A., de La Cueva, I., Dreizler, S., Endl, M., Giesers, B. and Jeffers, S.V., 2016. A terrestrial planet candidate in a temperate orbit around Proxima Centauri. nature, 536(7617), pp.437-440.

Notes

27. Anglada-Escudé, G., Amado, P.J., Barnes, J., Berdiñas, Z.M., Butler, R.P., Coleman, G.A., de La Cueva, I., Dreizler, S., Endl, M., Giesers, B. and Jeffers, S.V., 2016. A terrestrial planet candidate in a temperate orbit around Proxima Centauri. Nature, 536(7617), pp.437-440.

28. Optical Gravitational Lensing Experiment (OGLE) Collaboration Udalski A. 13 Szymański MK 13 Kubiak M. 13 Pietrzyński G. 13 14 Poleski R. 13 Soszyński I. 13 Wyrzykowski Ł. 15 Ulaczyk K. 13, 2011. Unbound or distant planetary mass population detected by gravitational microlensing. Nature, 473(7347), pp.349-352.

29. Miret-Roig, N., Bouy, H., Raymond, S.N., Tamura, M., Bertin, E., Barrado, D., Olivares, J., Galli, P.A., Cuillandre, J.C., Sarro, L.M. and Berihuete, A., 2022. A rich population of free-floating planets in the Upper Scorpius young stellar association. Nature Astronomy, 6(1), pp.89-97.

30. Pearson, S.G. and McCaughrean, M.J., 2023. Jupiter Mass Binary Objects in the Trapezium Cluster. *arXiv preprint arXiv:2310.01231*. URL: https://arxiv.org/pdf/2310.01231.pdf

31. Di Stefano, R., Berndtsson, J., Urquhart, R., Soria, R., Kashyap, V.L., Carmichael, T.W. and Imara, N., 2021. A possible planet candidate in an external galaxy detected through X-ray transit. Nature Astronomy, 5(12), pp.1297-1307.

6. Looking for the Signatures of Life

1. https://www.theguardian.com/film/2005/jun/17/sciencefictionfantasyandhorror.margaretatwood

2. Chela-Flores, J., 2012. The new science of astrobiology: from genesis of the living cell to evolution of intelligent behaviour in the universe (Vol. 3). Springer Science & Busi-ness Media.

3. Benner, S.A., 2010. Defining life. Astrobiology, 10(10), pp.1021-1030. https://doi.org/ 10.1089/ast.2010.0524.

4. Popa, R., 2004. Between necessity and probability: searching for the definition and origin of life. Springer Science & Business Media. https://doi.org/10.1007/ s11084-005-2042-z.

5. Greaves, J.S., Richards, A., Bains, W., Rimmer, P.B., Sagawa, H., Clements, D.L., Seager, S., Petkowski, J.J., Sousa-Silva, C., Ranjan, S. and Drabek-Maunder, E., 2021. Phosphine gas in the cloud decks of Venus. Nature Astronomy, 5(7), pp.655-664.

6. Snellen, I.A.G., Guzman-Ramirez, L., Hogerheijde, M.R., Hygate, A.P.S. and Van der Tak, F.F.S., 2020. Re-analysis of the 267 GHz ALMA observations of Venus-No statistically significant detection of phosphine. Astronomy & Astrophysics, 644, p.L2.

7. Ahrer, E.M., Stevenson, K.B., Mansfield, M., Moran, S.E., Brande, J., Morello, G., Murray, C.A., Nikolov, N.K., Petit Dit de la Roche, D.J., Schlawin, E. and Wheatley, J., 2023. Early Release Science of the exoplanet WASP-39b with JWST NIRCam. Nature, 614(7949), pp.653-658.

8. Madhusudhan, N., Sarkar, S., Constantinou, S., Holmberg, M., Piette, A. and Moses, I., 2023. Carbon-bearing Molecules in a Possible Hycean Atmosphere. arXiv preprint arXiv:2309.05566.

9. Schwieterman, E.W., Cockell, C.S. and Meadows, V.S., 2015. Nonphotosynthetic pigments as potential biosignatures. Astrobiology, 15(5), pp.341-361.

10. Board, S.S. and National Academies of Sciences, Engineering, and Medicine, 2019. An Astrobiology Strategy for the Search for Life in the Universe. National Academies Press.

Notes

11. Krissansen-Totton, J., Bergsman, D.S. and Catling, D.C., 2016. On detecting biospheres from chemical thermodynamic disequilibrium in planetary atmospheres. Astrobiology, 16(1), pp.39-67.

12. Marshall, S.M., Mathis, C., Carrick, E., Keenan, G., Cooper, G.J., Graham, H., Craven, M., Gromski, P.S., Moore, D.G., Walker, S.I. and Cronin, L., 2021. Identifying molecules as biosignatures with assembly theory and mass spectrometry. Nature communications, 12(1), p.3033.

13. McKay, D.S., Gibson Jr, E.K., Thomas-Keprta, K.L., Vali, H., Romanek, C.S., Clemett, S.J., Chillier, X.D., Maechling, C.R. and Zare, R.N., 1996. Search for past life on Mars: possible relic biogenic activity in Martian meteorite ALH84001. Science, 273(5277), pp.924-930.

14. Steele, A., Benning, L.G., Wirth, R., Schreiber, A., Araki, T., McCubbin, F.M., Fries, M.D., Nittler, L.R., Wang, J., Hallis, L.J. and Conrad, P.G., 2022. Organic synthesis associated with serpentinization and carbonation on early Mars. Science, 375(6577), pp.172-177.

15. Tarter, J.C., 2006. The evolution of life in the Universe: are we alone? Proceedings of the International Astronomical Union, 2(14), pp.14-29.

16. Dick, S.J., 1993. The search for extraterrestrial intelligence and the NASA High Resolution Microwave Survey (HRMS): Historical perspectives. Space science reviews, 64(1-2), pp.93-139.

17. Sheikh, S.Z., 2020. Nine axes of merit for technosignature searches. International Journal of Astrobiology, 19(3), pp.237-243.

18. Boyajian, T.S., Alonso, R., Ammerman, A., Armstrong, D., Ramos, A.A., Barkaoui, K., Beatty, T.G., Benkhaldoun, Z., Benni, P., Bentley, R.O. and Berdyugin, A., 2018. The first post-Kepler brightness dips of KIC 8462852. The Astrophysical Journal Letters, 853(1), p.L8.

19. Participants, N.A.S.A., 2018. NASA and the Search for Technosignatures: A Report from the NASA Technosignatures Workshop. arXiv preprint arXiv:1812.08681.

20. https://aleruriphd-axes-of-merit-aom-k1tpkw.streamlit.app/

21. Beatty, T.G., 2022. The detectability of nightside city lights on exoplanets. Monthly Notices of the Royal Astronomical Society, 513(2), pp.2652-2662.

22. Kardashev, N.S., 1964. Transmission of Information by Extraterrestrial Civilizations. Soviet Astronomy, Vol. 8, p. 217, 8, p.217.

23. BP (2020) Statistical Review of World Energy 72th edition.

24. Lemarchand, G., 1994. Detectability of extraterrestrial technological activities. The Columbus Optical SETI Observatory.–1992. URL: http://www.coseti.org/lemarch1.htm

25. Sagan, C. The Cosmic Connection: An Extraterrestrial Perspective; Cambridge University Press: Cambridge, UK, 1973; ISBN 13:978-0440133018.

26. Ruiz Rivera, A., 2015. Green traffic engineering techniques for current and next generation networks. URL: https://ro.uow.edu.au/cgi/viewcontent.cgi?article=5582&context=theses

27. Dyson, F.J., 1960. Search for artificial stellar sources of infrared radiation. Science, 131(3414), pp.1667-1668.

28. Timofeev, M.Y., Kardashev, N.S. and Promyslov, V.G., 2000. A search of the IRAS database for evidence of Dyson Spheres. Acta Astronautica, 46(10-12), pp.655-659.

29. Віщун, CC BY-SA 4.0 <https://commons.wikimedia.org/wiki/File:Dyson_Swarm_realistic_representation_cropped.jpg>

Notes

30. Griffith, R.L., Wright, J.T., Maldonado, J., Povich, M.S., Sigurdsson, S. and Mullan, B., 2015. The Ĝ infrared search for extraterrestrial civilizations with large energy supplies. III. The reddest extended sources in WISE. The Astrophysical Journal Supplement Series, 217(2), p.25.

31. Garrett, M.A., 2015. Application of the mid-IR radio correlation to the Ĝ sample and the search for advanced extraterrestrial civilisations. Astronomy & Astrophysics, 581, p.L5.

32. Loeb, A., 2021. Extraterrestrial: The first sign of intelligent life beyond earth. Houghton Mifflin.

33. Seligman, D. and Laughlin, G., 2020. Evidence that 1I/2017 U1 ('Oumuamua) was composed of molecular hydrogen ice. The Astrophysical Journal Letters, 896(1), p.L8.

34. Guzik, P., Drahus, M., Rusek, K., Waniak, W., Cannizzaro, G. and Pastor-Marazuela, I., 2020. Initial characterization of interstellar comet 2I/Borisov. Nature Astronomy, 4(1), pp.53-57.

35. Seager, S., Petkowski, J.J., Huang, J., Zhan, Z., Ravela, S. and Bains, W., 2023. Fully fluorinated non-carbon compounds NF3 and SF6 as ideal technosignature gases. Scientific Reports, 13(1), p.13576.

36. Wright, J.T., Kanodia, S. and Lubar, E., 2018. How much SETI has been done? Finding needles in the n-dimensional cosmic haystack. The Astronomical Journal, 156(6), p.260.

37. Gray, R.H., 2012. *The elusive wow: searching for extraterrestrial intelligence.* Palmer Square Press.

38. https://www.sparkfun.com/news/4664

39. Paris, A. and Davies, E., 2015. Hydrogen Clouds from Comets 266/P Christensen and P/2008 Y2 (Gibbs) are Candidates for the Source of the 1977 "WOW" Signal. Journal of the Washington Academy of Sciences, 101(4), pp.25-32.

40. Whitmire, Daniel P., and David P. Wright. "Nuclear waste spectrum as evidence of technological extraterrestrial civilizations. Icarus 42, no. 1 (1980): 149-156.

41. Przybylski, A., 1961. HD 101065—a G 0 Star with High Metal Content. Nature, 189(4766), pp.739-739.

42. Xiangyuan Ma, P., Ng, C., Rizk, L., Croft, S., Siemion, A.P., Brzycki, B., Czech, D., Drew, J., Gajjar, V., Hoang, J. and Isaacson, H., 2023. A deep-learning search for technosignatures of 820 nearby stars. arXiv e-prints, pp.arXiv-2301.

43. Petroff, E., Keane, E.F., Barr, E.D., Reynolds, J.E., Sarkissian, J., Edwards, P.G., Stevens, J., Brem, C., Jameson, A., Burke-Spolaor, S. and Johnston, S., 2015. Identifying the source of perytons at the Parkes radio telescope. Monthly Notices of the Royal Astronomical Society, 451(4), pp.3933-3940.

44. Lesnikowski, A., Bickel, V.T. and Angerhausen, D., 2020. Unsupervised distribution learning for lunar surface anomaly detection. arXiv preprint arXiv:2001.04634.

45. Liu, C., 2016. Death's end (Vol. 3). The three-body problem trilogy. Macmillan.

46. Stancil, D.D., Adamson, P., Alania, M., Aliaga, L., Andrews, M., Del Castillo, C.A., Bagby, L., Bazo Alba, J.L., Bodek, A., Boehnlein, D. and Bradford, R., 2012. Demonstration of communication using neutrinos. Modern Physics Letters A, 27(12), p.1250077.

47. IceCube Collaboration* †, Abbasi, R., Ackermann, M., Adams, J., Aguilar, J.A., Ahlers, M., Ahrens, M., Alameddine, J.M., Alves Jr, A.A., Amin, N.M. and Andeen, K., 2023. Observation of high-energy neutrinos from the Galactic plane. Science, 380(6652), pp.1338-1343.

Notes

7. The Great Silence – Are we alone?

1. https://www.nasa.gov/wp-content/uploads/2015/01/archaeology_anthropology_and_in terstellar_communication_tagged.pdf
2. Garber, S.J., 1999. Searching for good science-the cancellation of NASA's SETI program. Journal of the British Interplanetary Society, 52(1), pp.3-12.
3. Worden, S.P., Drew, J., Siemion, A., Werthimer, D., DeBoer, D., Croft, S., MacMahon, D., Lebofsky, M., Isaacson, H., Hickish, J. and Price, D., 2017. Breakthrough listen–a new search for life in the universe. Acta Astronautica, 139, pp.98-101.
4. https://breakthroughinitiatives.org/initiative/1
5. van Dokkum, P., 2023. An exciting era of exploration. Nature Astronomy, 7(5), pp.514-515.
6. Lineweaver, C.H., 2001. An estimate of the age distribution of terrestrial planets in the universe: quantifying metallicity as a selection effect. Icarus, 151(2), pp.307-313.
7. MacKenzie, S.M., Neveu, M., Davila, A.F., Lunine, J.I., Craft, K.L., Cable, M.L., Phillips-Lander, C.M., Hofgartner, J.D., Eigenbrode, J.L., Waite, J.H. and Glein, C.R., 2021. The Enceladus Orbilander mission concept: Balancing return and resources in the search for life. The Planetary Science Journal, 2(2), p.77.
8. Neumann, J.V., 1966. Theory of self-reproducing automata. Edited by Arthur W. Burks.
9. Cirkovic, M.M., 2009. Fermi's paradox-The last challenge for copernicanism?. arXiv preprint arXiv:0907.3432.
10. Webb, S., 2002. If the universe is teeming with aliens... where is everybody?: fifty solutions to the Fermi paradox and the problem of extraterrestrial life (p. 112). New York, NY: Copernicus Books.
11. Webb, S., 2015. If the universe is teeming with aliens... where is everybody?: seventy-five solutions to the fermi paradox and the problem of extraterrestrial life. Heidelberg: Springer International Publishing.
12. Robitaille, T.P. and Whitney, B.A., 2010. The present-day star formation rate of the milky way determined from spitzer-detected young stellar objects. The Astrophysical Journal Letters, 710(1), p.L11.
13. Nazari-Sharabian, M., Aghababaei, M., Karakouzian, M. and Karami, M., 2020. Water on Mars—a literature review. Galaxies, 8(2), p.40.
14. Martínez, G. and Renno, N.O., 2013. Water and brines on Mars: current evidence and implications for MSL. Space Science Reviews, 175, pp.29-51.
15. https://astrobiology.com/2016/05/hunting-for-hidden-life-on-worlds-orbiting-old-red-stars.html
16. Brown, David W..The Mission: A True Story. United States: Custom House, 2021.
17. Saur, J., Duling, S., Roth, L., Jia, X., Strobel, D.F., Feldman, P.D., Christensen, U.R., Retherford, K.D., McGrath, M.A., Musacchio, F. and Wennmacher, A., 2015. The search for a subsurface ocean in Ganymede with Hubble Space Telescope observations of its auroral ovals. Journal of Geophysical Research: Space Physics, 120(3), pp.1715-1737.
18. McKay, C.P., 2016. Titan as the abode of life. Life, 6(1), p.8.
19. Hansen, C.J., Castillo-Rogez, J., Grundy, W., Hofgartner, J.D., Martin, E.S., Mitchell, K., Nimmo, F., Nordheim, T.A., Paty, C., Quick, L.C. and Roberts, J.H., 2021. Triton: fascinating moon, likely ocean world, compelling destination!. The Planetary Science Journal, 2(4), p.137.

Notes

20. https://astrobites.org/2014/01/17/how-to-keep-warm-outside-the-habitable-zone/
21. Spohn, T. and Schubert, G., 2003. Oceans in the icy Galilean satellites of Jupiter? Icarus, 161(2), pp.456-467.
22. https://www.youtube.com/channel/UCGHZpIpAWJQ-Jy_CeCdXhMA
23. Mann, J. and Patterson, E.M., 2013. Tool use by aquatic animals. Philosophical Transactions of the Royal Society B: Biological Sciences, 368(1630), p.20120424.
24. Mercado III, E., Green, S.R. and Schneider, J.N., 2008. Understanding auditory distance estimation by humpback whales: a computational approach. Behavioural processes, 77(2), pp.231-242.
25. https://www.universetoday.com/162980/the-space-station-is-getting-gigabit-internet/
26. Balbi, A. and Frank, A., 2023. The Oxygen Bottleneck for Technospheres. arXiv preprint arXiv:2308.01160.
27. Olga, L. (2022) *Who Was Vasili Arkhipov?: A Biography and Story of the Russian That Saved the World from Nuclear War in 1962.* Amazon Digital Services LLC - Kdp Print Us.
28. https://www.youtube.com/watch?v=aspMV6ERqpo
29. Hapgood, M., 2012. Prepare for the coming space weather storm. Nature, 484(7394), pp.311-313.
30. https://www.spacecentre.nz/resources/tools/drake-equation-calculator.html
31. Seager, S., 2018. The search for habitable planets with biosignature gases framed by a 'Biosignature Drake Equation'. International Journal of Astrobiology, 17(4), pp.294-302.
32. https://www.zooniverse.org/projects/nora-dot-eisner/planet-hunters-tess/about/research
33. Bekker, A., Holland, H.D., Wang, P.L., Rumble Iii, D., Stein, H.J., Hannah, J.L., Coetzee, L.L. and Beukes, N.J., 2004. Dating the rise of atmospheric oxygen. Nature, 427(6970), pp.117-120.
34. Smith, H.B. and Mathis, C., 2022. The Futility of Exoplanet Biosignatures. arXiv preprint arXiv:2205.07921.
35. https://iaaspace.org/wp-content/uploads/iaa/Scientific%20Activity/setideclaration.pdf
36. Chon-Torres, O.A., Chela-Flores, J., Dunér, D., Persson, E., Milligan, T., Martínez-Frías, J., Losch, A., Pryor, A. and Murga-Moreno, C.A., 2024. Astrobiocentrism: reflections on challenges in the transition to a vision of life and humanity in space. International Journal of Astrobiology, 23, p.e6.
37. Chon-Torres, O.A., 2018. Astrobioethics. International Journal of Astrobiology, 17(1), pp.51-56.

Final Remarks

1. Adams, D., 1995. The Hitch Hiker's Guide to the Galaxy Omnibus. Random House.
2. Rampelotto, P.H., 2013. Extremophiles and extreme environments. Life, 3(3), pp.482-485.
3. Iglesias-Groth, S. and Marin-Dobrincic, M., 2023. A rich molecular chemistry in the gas of the IC 348 star cluster of the Perseus Molecular Cloud. Monthly Notices of the Royal Astronomical Society, 521(2), pp.2248-2269.
4. https://www.space.com/finding-life-outside-solar-system-in-25-years
5. Ellery, A., 2004. Rare Earth–why complex life is uncommon in the universe Peter Ward and Donald Brownlee Copernicus Publishers, New York, USA (2000) 335 pages· ISBN 0-387-95289-6. International Journal of Astrobiology, 3(1), pp.70-70.

Notes

6. Prantzos, N., 2008. On the "galactic habitable zone". Strategies of Life Detection, pp.313-322.
7. Horner, J. and Jones, B.W., 2010. Jupiter: friend or foe? An answer. Astronomy & Geophysics, 51(6), pp.6-16.